SCOPE OF SELECTIVE HETEROCYCLES FROM ORGANIC AND PHARMACEUTICAL PERSPECTIVE

Edited by **Ravi Varala**

Scope of Selective Heterocycles from Organic and Pharmaceutical Perspective

http://dx.doi.org/10.5772/60890
Edited by Ravi Varala

Contributors

Someshwar Pola, Premlata Kumari, Amit Patel, Joaquín M. Campos, Verónica Gómez-Pérez, Santiago Castanys, Francisco Gamarro, Raghunath Baban Toche, Purna Bhavnari, Ravi Varala, Rodica Mihaela Dinica, Ioana Otilia Ghinea

CBS, Edition **2017**

Published by InTech

Janeza Trdine 9, 51000 Rijeka, Croatia

Publishing Process Manager Iva Simcic

Technical Editor InTech DTP team

Cover Designer InTech Design Team

Additional hard copies can be obtained from orders@intechopen.com

Scope of Selective Heterocycles from Organic and Pharmaceutical Perspective, Edited by Ravi Varala
p. cm.
Print ISBN 978-953-51-2503-7
Online ISBN 978-953-51-2504-4

INTECH

open science | open minds

Contents

Preface

Scope of Selective Heterocycles from Organic and Pharmaceutical Perspective is a compilation of bioactive-chosen heterocyclic scaffolds intended for postgraduates, research scholars, pharmaceutical scientists, and others interested in an appreciation of the title subject. It is an edited book and is not comprehensive as well in the mentioned field. Few synthetic strategies along with bioactivity are presented, and some limitations were raised in order to arouse curiosity of the reader. This book includes six chapters, written by international experts, and maximum care is taken to assemble them in an order so that the reader's interest is generated. Compiling this book was really a learning experience for me, through which I improved my knowledge about book writing and editing.

I am thankful to InTech Publishers, who gave me this wonderful opportunity to dare to edit my first book. I am thankful to Prof. Appalanaidu, Prof. Y. Rajeshwer Rao, and Prof. Sirasani Satyanarayana, for their courtesy and affection toward me. I ought to thank FAPESP, Brazil, for funding my visiting scientist position for the project work mentioned in one of the book chapters (Grant No. 2014/25784-7). Finally, I thank RGUKT, Basar, for supporting me in all ways.

Dr. Ravi Varala
Head, Department of Chemistry
Rajiv Gandhi University of Knowledge Technologies
Basar, Mudhole (M), Adilabad
Telangana State, India

Chapter 1

Significance of Thiazole-based Heterocycles for Bioactive Systems

Someshwar Pola

Additional information is available at the end of the chapter

http://dx.doi.org/10.5772/62077

Abstract

Monocyclic and Bicyclic aromatic heterocycles such as imidazoles, thiazoles, thiadiazoles, oxazoles, oxadiazoles quinazolines, indoles, benzimidazoles, purines pyrido[4,3-d]pyrimidines, thiazolo[5,4-d]pyrimidines, thiazolo[4,5-d]pyrimidines, oxazolo[5,4-d]pyrimidines and thieno[2,3-d]pyrimidines are renowned pharmacophores in drug discovery. These special structures are well explained and exemplified in chemical compound libraries. In this chapter, several types of thiazole based heterocyclic scaffolds such as monocyclic or bicyclic systems synthesis and their biological activities studies are presented, which are not frequently present in books and reviews. We mention the first importance of synthetic route of various thiazole based compounds and their applications in medicinal chemistry in this chapter.

Keywords: Thiazole, privileged structures, thiazolopyridine, thiazolopyrimidines

1. Introduction

Currently, the whole pharmaceutical industry is encountered with the challenge of enhancing work rate and advancement. The key obstacles are the increasing expenses of exploration and expansion and a concurrent deteriorating amount of new chemical entities (NCEs). The source of this modernism shortfall is not the biology. Interpreting of the human genome has directed to a prosperity of drug targets. With the addition of more than 35,000 human genes, the hypothesis is that at least 2,000 are significantly tangled in the occurrence and progress of the illness. Moreover, since each of these genes is associated with the usefulness of between five and ten proteins, the deduction is that their potency be 5,000 – 10,000 aims for innovative drugs [1,2]. Even though the positive outline of protein therapeutics and the aptitude of gene therapy, key pharmaceutical establishments are even focused on research and growth of small molecular mass compounds. Therefore, the challenge is to choose the greatest drugable objectives and formulate the conforming drug-like molecules.These materials are not only relative to the mark but also have precise pharmacokinetic and toxicological properties, which was allowed to be established as a drug. Medicinal chemistry as a scientific discipline has introduced several new techniques over the last few years to the rapidity of the drug discovery process, such as combinatorial chemistry, microwave-assisted organic synthesis, and high - output refinement [3]. Despite the stable rise in R & D, the total number of NCE successes in the market has reduced fundamentally. It appears clearly that choosing the suitable molecules to synthesize is one of the most difficult queries. It has been projected that the sum of potential compounds with molecular weight of lower than 500 Da is 10^{200}, where only 10^{60} may retain drug-like applications. The percentage of molecules prepared untill today has been projected as one part in 10^{58} or approximately the fraction of the mass of the proton to the mass of the sun. The concern is, therefore, the selection of new molecules from this vast universe that have the potential to be biologically active [4]. To build a new drug discovery mission and to discover the bioactive compounds, various possibilities are offered. Triumphs can be achieved via a virtual screening method or can be simulated from technical or manifest literature. Most often than not, drug innovation projects start with a high quantity screening operation of commercially accessible compound collections besides targeting curiosity. It became clear in recent years that combinatorial libraries are not distinct enough. As the core attention of the Laboratory of Medicinal Chemistry showed in the synthesis and biological evaluation of bicyclic aromatic heterocycles [5], it is scrutinized that the number of accessible bicyclics heterocycles is principally restricted to a well-known nitrogen enclosing compounds, such as pyrimidines, thioazoles, coumarins, thiozlopyridines and benzothiazole (Figure 1).

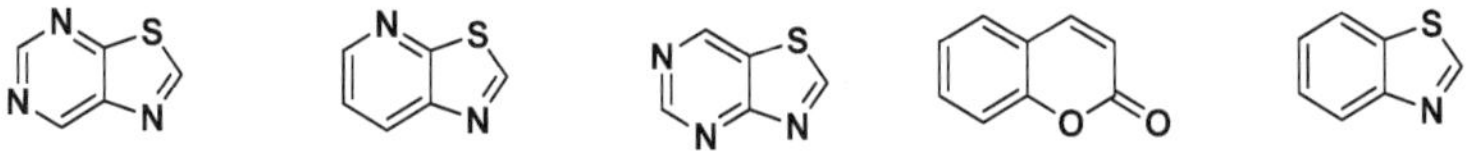

Figure 1. Examples of privileged structures

In vision of the significance of thiazoles and their derivatives, numerous approaches for its synthesis were developed by various groups such as Hantzsch [6], Tchernic [7], Cook-Heilborn and Gabriel [8].

A thiazole ring system originates naturally in the crucial water soluble vitamin thiamin, also known as Vitamin B1, which supports the discharge of energy from carbohydrates through the course of metabolism. The occurrence of thiazole ring in vitamin B1 and its coenzyme play a significant role in the decarboxylation of α-keto acids and as an electron sink, respectively [9]. It also assist in the regular operational of the nervous system through its character in the synthesis of acetylcholine, a neurotransmitter.

Thiazole ring system appears in the bacitracin and penicillin antibiotics and various synthetic drugs. Synthetic drugs belonging to the thiazole family consist of the antimicrobial agents acinitrazole **(1)** and sulfathiazole [10], **(2)** antibiotic penicillin [11], **(3)** antidepressant pramipexole [12], **(4)** antineoplastic agents Bleomycin **(5)** and Tiazofurin [13], (6) anti-HIV drug Ritonavir [14], **(7)** the antiasthmatic drug cinalukast [15], **(8)** antiulcer agent Nizatidine [16] **(9)**. Additionally, extensively used thiazole derivatives are the non-steroidal immunomodulatory drug Fanetizole [17] **(10)** and anti-inflammatory drug Meloxicam [18] **(11)**. Thiazole derivatives with polyoxygenated phenyl module have exhibited encouraging anti-fungal activity [19]. Thiazoles found from microbial, and marine ancestries reveal antitumor and antiviral activities. Thiazole is recognized as ligand of estrogen receptors [20] and also as unique kind of antagonists for adenosine receptors [21].

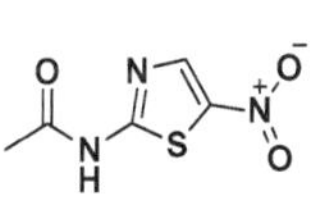
Acinitrazole (1)

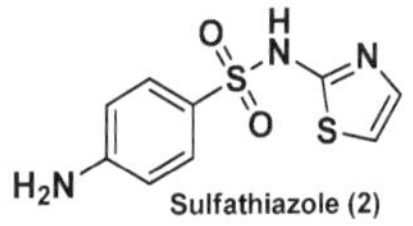
Sulfathiazole (2)

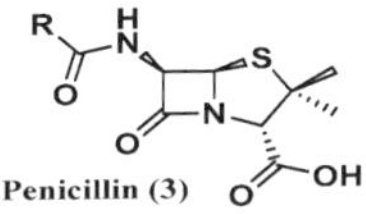
Penicillin (3)

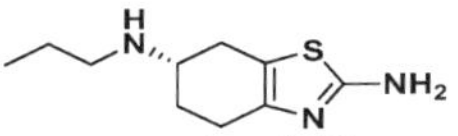
pramipexole (4)

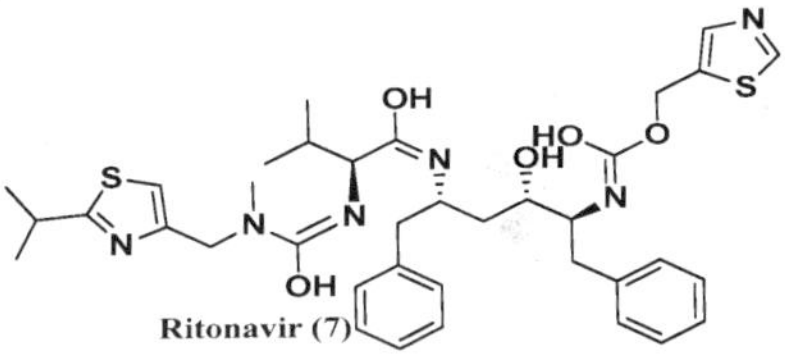
Ritonavir (7)

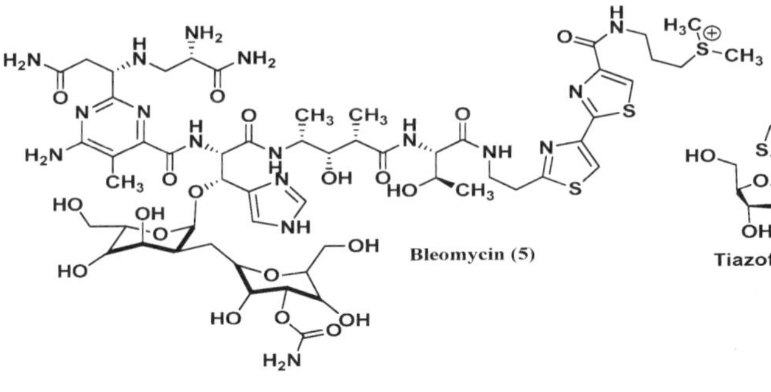
Bleomycin (5)

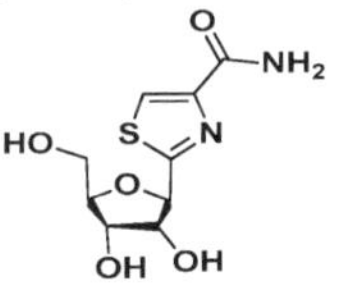
Tiazofurin (6)

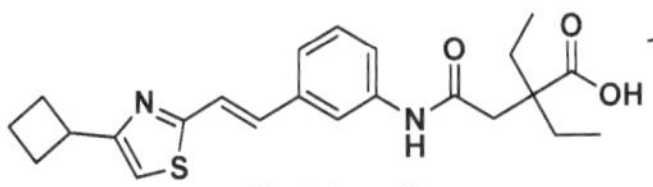
Cinalukast (8)

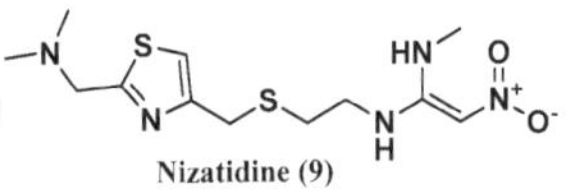
Nizatidine (9)

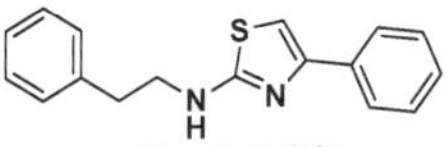
Fanetizole (10)

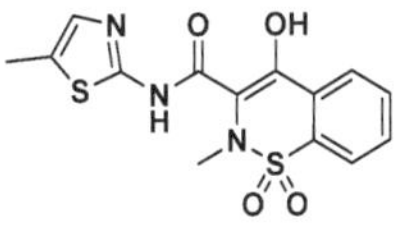
Meloxicam (11)

Other substantial thiazoles take account of essential dyes and fungicides or nematicide, Tricyclazole **12**, Thiabendazole **13**, and Thifluzamide **14** are promoted for the switch of several agricultural pests [22,23]. Primuline yellow **15** and Rhodanine red **16** dyes are some of the best models of thiazole moiety containing dyes [24,25]. Numerous thiazoles are flavor materials and also originate in roasted peanuts. They materialized in foods by the exploit of sulfur-containing amino acids interacting with carbohydrates. Thiazoles are surrounded by some significant heterocyclic compounds that give the flavor of fermented coffee [26].

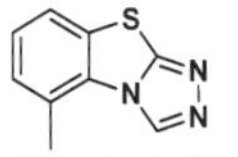

Tricyclazole (12)

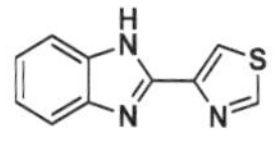

Thiabendazole (13)

Thifluzamide (14)

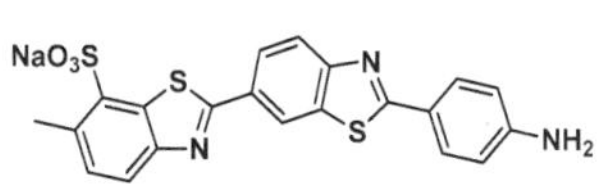

Primuline yellow dye (15)

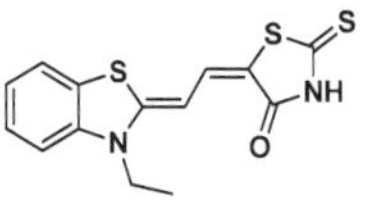

Rhodanine red dye (16)

The exhilarating outcomes of the 2,4-disubstituted thiazoles as a unique class of Src Homology 2 (SH2) inhibitors for the behavior of osteoporosis and breast cancer have also been reported [27]. Selection of the 2,4-disubstituted thiazoles as concealed pharmacophores for diacylhydrazine of SC-51089, a prospective PGE2 antagonist have also been described [28]. With these results, the thiazole ring system proves to be a well-known structural motif that originate in several pharmaceutical agents and natural products extracted from various plants and marine systems.

2. Structure of Thiazole

The structure of thiazole is reflected as the resonance amalgam of the subsequent resonating structures (Figure 1). However, some of the resonating structures are also probable with the contribution of d-orbitals of the sulfur atom.

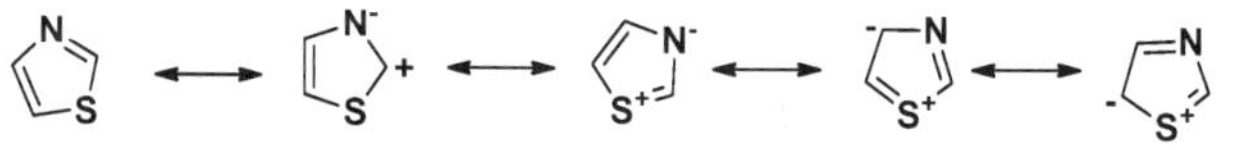

Figure 2. Resonating Structures of Thiazole

The p-bond orders quantified by molecular orbital methods have specified thiazole molecule to be aromatic with some dienic nature. Localization energies have projected reducing order of the nucleophilic reactivities following the order: 2 > 5 > 4 and the electrophilic reactivities as: 5 > 2 > 4. Three hydrogen atoms present in the thiazole are anticipated to have the order of acidity as H2 >> H5 > H4.

3. Synthesis of Thiazole

In the assessment of the significance of thiazoles and their derivatives, numerous techniques for the synthesis of thiazole derivatives were established by various research groups such as Hantzsch [6], Tchernic [7], Cook-Heilborn and Gabriel [8]. Lately, thiazole derivatives were generated in the presence of various catalysts such as ammonium-12-molybdophosphate [29], cyclodextrins [30], iodine [31] and silica chloride [32] in organic solvents at higher temperature and solvents such as 1-methyl-2-pyrrolidinone [33], with the use of a microwave [34]. Numerous procedures for the synthesis of thiazole compounds are accessible, which can be categorized into the part structures demonstrated below. The earliest of these structures is observed to be the most significant and highly flexible of all the thiazole formation techniques. With a workable and first reactants, it approves alkyl, aryl, aralkyl or heterocycles to be taken in any one of the 2-, 3-, 4- or 5-carbons of the thiazole ring. This technique, better acknowledged by the name of the German chemist Hantzsch, who invented it in 1887, contains the condensation of a compound bearing the two heteroatoms on the same carbon with a compound attached one halogen and one carbonyl function on two adjacent carbon atoms. A boundless diversity of compounds may assist as nucleophilic reagent in this reaction, such as thiourea, thioamide, ammonium thiocarbamate or dithiocarbamate and its derivatives [35].

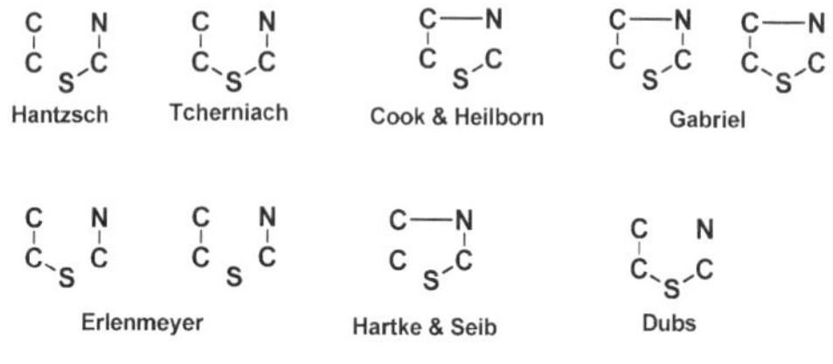

3.1. Synthesis from α-halocarbonyl compounds (Type Ia): Hantzsch's synthesis.

First designated in 1887 by Hantzsch, the cyclization of α-halo carbonyl compounds by a wide diversity of reactants attached to the N-C-S portion of the ring is the most extensively popular process for formation of thiazoles.

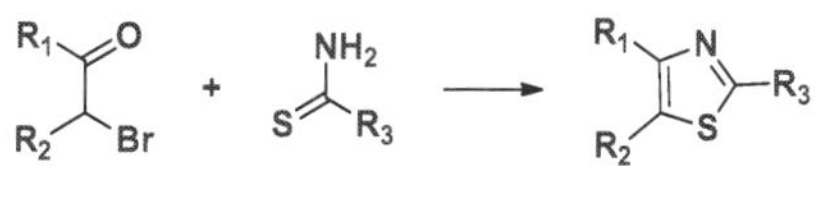

3.1.1. Reactions with Thioamides

3.1.1.1. Chloroacetaldehyde and derivatives

Thiazole ready to obtain by condensing thioformamide and chloroacetaldehyde [36,37].

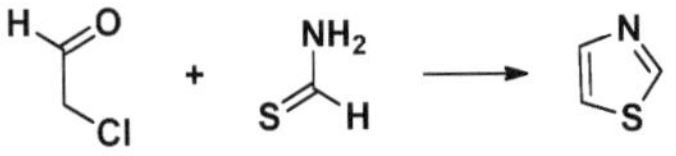

3.1.1.2. Condensation with higher thioamides (2,4-Disubstituted and 2,4,5-trisubstituted thiazoles)

The reaction between thioamide and various α-halocarbonyl compounds has been utilized broadly, and numerous thiazoles with alkyl, aryl, arylalkyl or heteroaryl of several functional groups at 2-, 4- or 5-positions have been published.

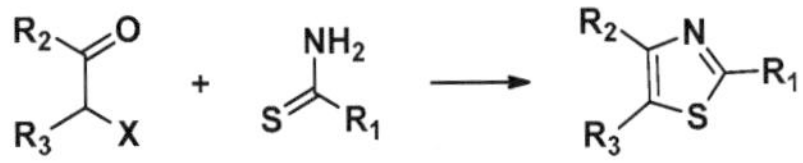

3.1.2. Reactions with N-substituted Thiourea

3.1.2.1. N-monosubstituted thioureas

The 2-monosubstituted or disubstituted aminothiazoles obtained reaction between Halo carbonyl *and* N-substituted thiourea compounds [38].

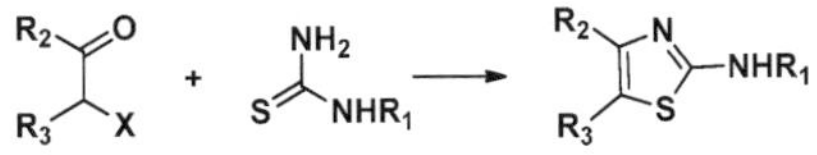

3.1.3. Reaction with salts and esters of thiocarbamic acid: 2-hydroxy thiazoles and derivatives

This technique, originated by Marchesini [39,40], in 1893 involves the condensation of a α-halocarbonyl compound with ammonium thiocarbamate to give 2-hydroxythiazole derivatives.

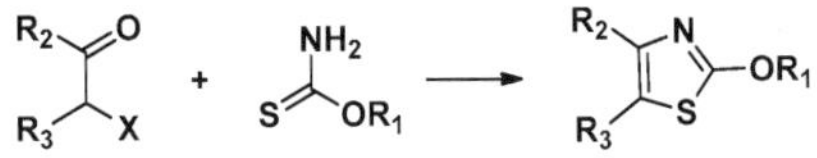

3.2. Thiazoles formation from reorganization of the α-thiocyanatoketones

The simple cyclic reaction of α-thiocyanatoketones in aqueous acid concentrated sulfuric acid in acetic acid, and water or alkaline solution gives to 2-hydroxy thiazoles after dilution in water. These reactions can be conceded out for various hours at room temperature or by refluxing for 1 or 2 hrs on a water bath [41-45].

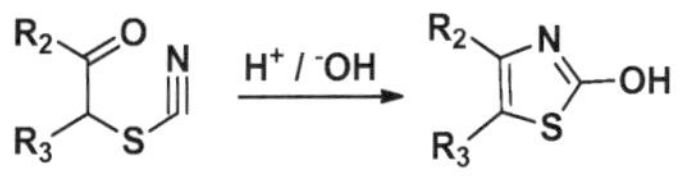

α-Thiocyanatoacetophenone reacts thioacid to yield 2-mercapto-4-phenyl thiazole.

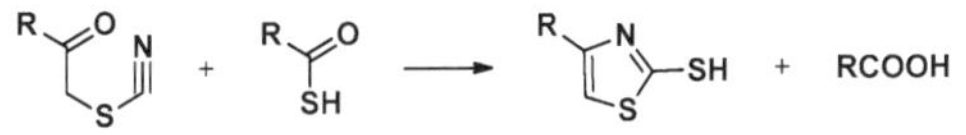

α-Thiocyanatoketones highly react with alkyl amine or ammonium chloride to provide their *N*-substituted derivatives or 2-aminothiazoles [46].

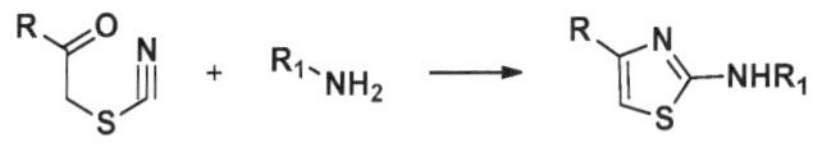

3.3. Thiazoles from α-aminonitriles (Cook-Heilbron's synthesis) (Type-II)

This category of synthesis, which was examined by Cook, Heilbron [47-49] give 5-aminothiazoles differently substituted in the 2-position by reacting with an aminonitrile with salts and esters of dithioacids, carbon oxysulfide, carbon disulfide, and isothiocyanates under remarkably very mild conditions.

3.3.1. Carbon disulfide: 2-mercapto-5-aminothiazole derivatives

Carbon disulfide freely responds with α-aminonitriles giving 2-mercapto-5-amino thiazoles [50,51], which can be transformed into 5-amino thiazoles unsubstituted in the 2-position.

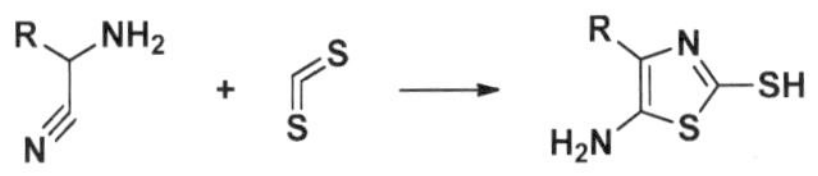

3.3.2. Esters and salts of dithioacids: 5-aminothiazole compounds and related condensations

By reducing the salts or the esters of both dithioformic and dithiophenacetic acids with α - aminonitriles, 5-aminothiazoles were achieved in better yields [52]. These reactions have agreed in aqueous ethereal solution at ambient temperature.

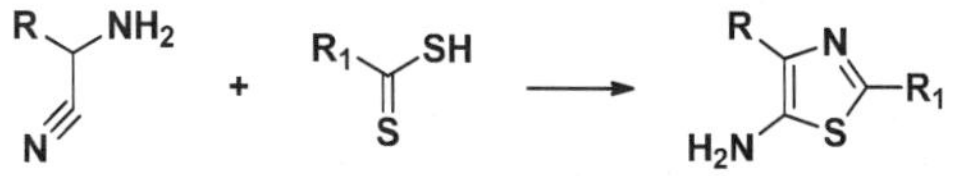

3.4. Thiazoles from acylaminocarbonyl compounds and phosphorus pentasulfide and related condensation (Gabriel's synthesis) (Type III)

This reaction was originally designated by Gabriel [53] in 1910 phosphorus pentasulfide reacted with acylaminoketone (showed in below reaction) an equimolecular quantity to yield 2-phenyl-5-alkyl-thiazole. The reaction is analogous to the synthesis of additional five-membered oxygen and sulfur holding rings from 1,4-dicarbonyl compounds.

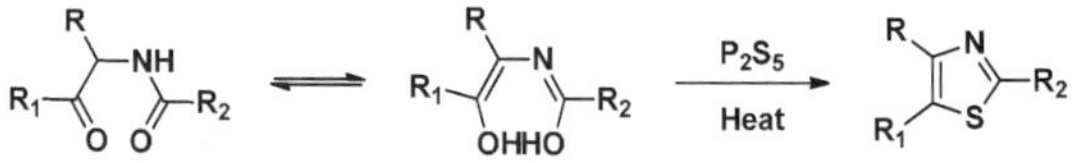

3.5. Thiazoles from nitriles and α-mercaptoketones: 2,4-disubstituted and 2,4,5-trisubstituted derivatives

Also, α-halocarbonyl compounds and α-mercaptoketones react with nitriles and aldehyde oximes in the presence of an acid as catalyzed reaction for the synthesis of thiazoles.

3.5.1. 2,4,5-Trisubstituted thiazoles from α-mercaptoketones and nitriles

Miyatake and Yashikawa synthesized numerous 2,4,5-trisubstituted thiazoles and gave low yield (16 to 40%) by the interaction of α-mercaptoketones on nitriles. Asinger and Thiel [54] utilized an aldehyde and ammonia as an alternative for nitrile.

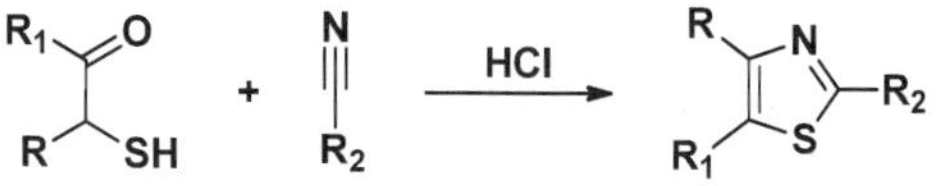

R, R_1, & R_2 are different substituents

3.5.2. 2,4-Diaminothiazole derivatives from α-halonitriles and thiourea

α-Halonitrile can substitute α-halogenocarbonyl compounds in the Hantzsch's synthesis [55-57], thus, the reaction of thiourea with a α-halonitrile in refluxing alcohol provides 2,4-diaminothiazoles.

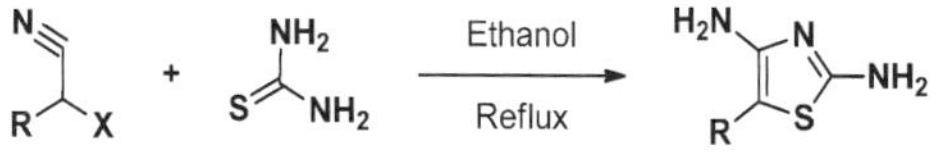

3.6. Thiazoles from Vinyl Bromide

Thiazoles holding a variability of substituents such as aliphatic, aromatic, heterocyclic, or alkenyl groups can be synthesized by an intramolecular nucleophilic substitution reaction of *N*-(2-bromoprop-2-enyl)thioamides [58]. This vinylic substitution technique would afford an exclusive synthetic method for a range of heterocycles.

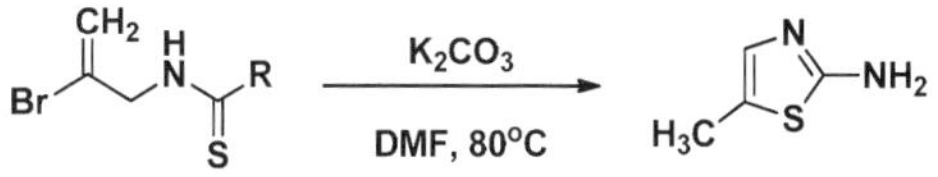

3.7. Synthesis of 2,4-disubstituted-5-acetoxythiazoles

From the viable existing methyl benzoate derivatives and with racemic phenyl glycine, a range of 2,4- disubstituted-5-acetoxythiazoles obtained in worthy to reasonable yields exhausting the succeeding scheme [59]. Due to the excellent thermal stability of the thiazole nucleus, the polymers integrating thiazole ring protocol have also been prepared.

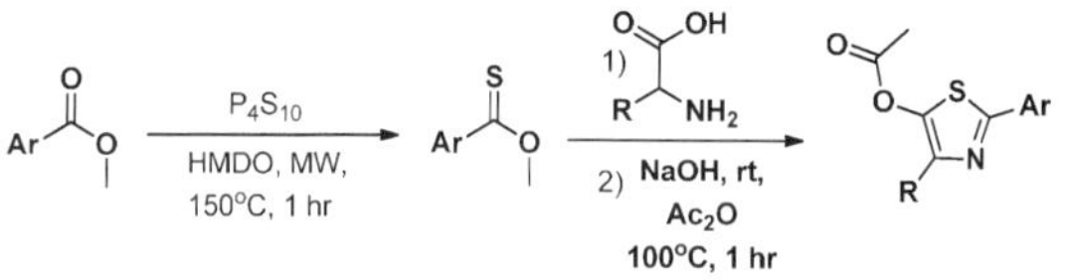

4. Biological importance of thiazoles

Thiazole moiety-containing compounds invention present an extensive range of applications in medicinal chemistry such as antibiotics, bacteriostatics, CNS regulants to high selling diuretics [60-64]. Thiazole framework has established wide application in drug growth for the treatment of hypertension [65], inflammation [66] and HIV infections [67]. Aminothiazoles are famous for being ligands of estrogen receptors [68] as well as a innovative type of adenosine receptor antagonists [69]. Other equivalents are utilized as fungicides, inhibiting *in vivo* progress of Xanthomonas, as a component of herbicides or as schistosomicidal and anthelmintic drugs [70].

Sherif. et al. [71] syntheses of two series of compounds that is thiazolylantipyrines and thiadiazolylantipyrines, in which thiazolylantipyrine series exhibits better antibacterial potencies than the thiadiazolylantipyrine series of compounds. In thiazolylantipyrine series compounds **17 – 19** are well thought-out to be the better active antimicrobial members recognized in this study with a broad spectrum of antibacterial activity against both Gram positive and Gram negative bacteria.

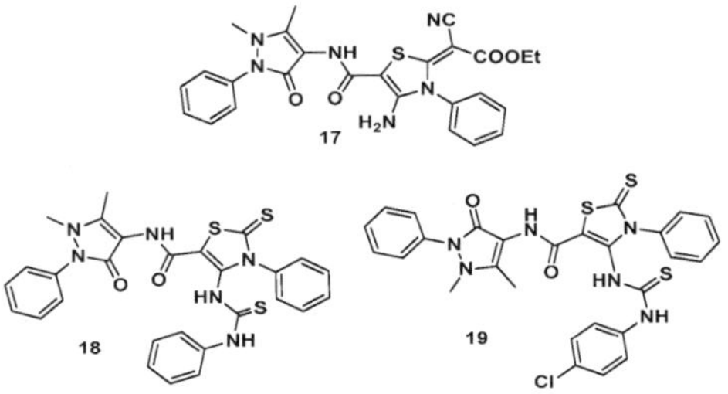

17 18 19

Zablotskaya A et al. [72] prepared trimethylsilyl ethers of different hydroxyl group bearing thiazole compounds. All the compounds examined possess antihypoxic properties and extend the life of mice under conditions of hypoxia by 20-78%. The silylated and unsilylated derivatives in the preponderance of circumstances show antihypoxic activity.

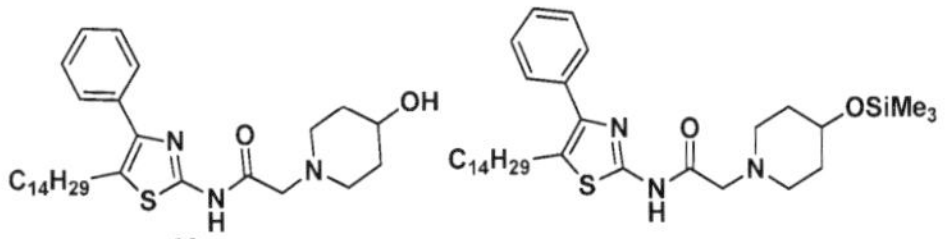

20 21

Dae-Kee K et al. [73] produced a set of 5-(pyridin-2-yl)thiazoles enclosing a *para* or *meta*-carboxamide or carbonitrile-substituted phenylmethylamino moiety at the 2-position of the thiazole ring and was estimated for activating receptor-like kinase 5 (ALK5) inhibitory activity in cell-based luciferase publisher assays.

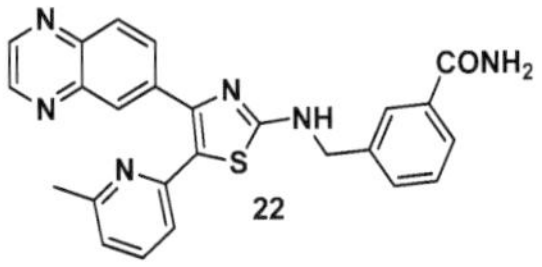

22

Rajan S G et al. [74] designed and synthesized a sequence of 2-(2,4-disubstituted-thiazole-5-yl)-3-aryl-3H-quinazoline-4-one **23** compounds. Synthesized molecules were estimated for their inhibitory activity in the course of record factors, nuclear factor-kB (NF-kB) and activating factor (AP-1) interceded transcriptional activation in a cell line based *in vitro* assay as well as for their anti-inflammatory activity *in vivo* model of severe inflammation.

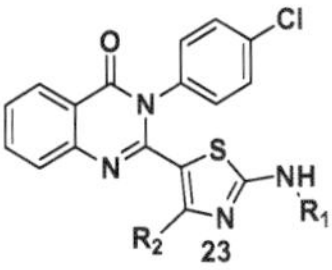

23

Johan et al. [75] synthesized a unique sequence for Aurora kinase inhibitors enclosing thiazole moiety (SNS-314, **24**). Also, key SAR as well as essential binding elements has been explained.

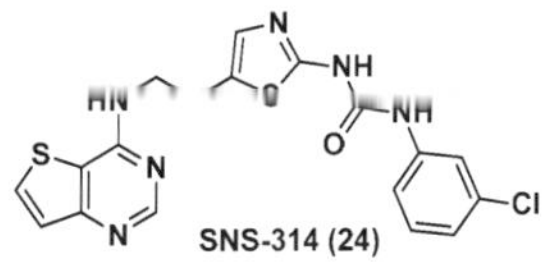

SNS-314 (24)

HI El-Subbagh et al. [76] synthesized a sequence of 2,4-disubstituted thiazole compounds containing *N*-n-butyl or *N*-cyclohexyl thioureido synthon at position-2 and N-substituted thiosemicarbazone moiety **25** at position-4 and verified for antitumor activity. All of the established derivatives revealed antineoplastic activity at concentrations less than 10^2 μM.

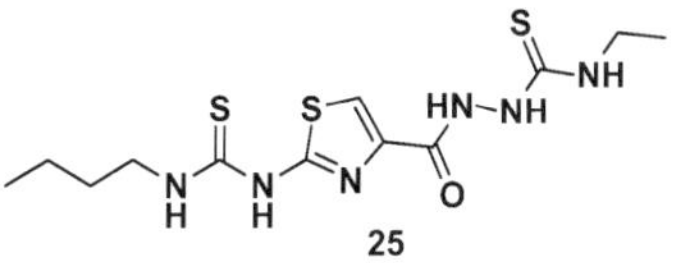

25

The unique model of a thiazole in the best 200 drugs citations is cefdinir **26** (Omnicef), a semi-synthetic third generation cephalosporin that is controlled orally and has a stretched antibacterial activity in contrast to both gram-positive and gram-negative bacteria. The key feature of cefdinir is that it exhibits outstanding activity against **Staphylococcus** species [77]. The thiazole ring in cefdinir reveals that the heterocyclic structure in a drug does not only affect its pharmacodynamic properties but can also affect its kinetics. It is hypothesized that the digestive tract iron (II) ions form chelate complexes with the oxime nitrogen atom and thiazole ring and, therefore, decrease the bioavailability of cefdinir [77].

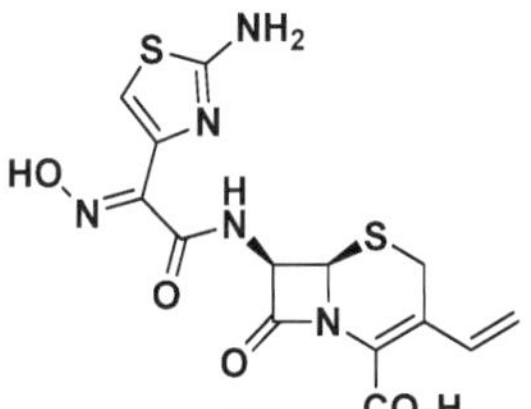

26

The HIV-1 protease inhibitor ritonavir [78] (Norvir **7**) contains two different substituted thiazole rings, which are presented at the advanced steps in the synthesis of this peptidomimetic antiviral compound. Remarkably, ritonavir is a consequence of advanced enhancements on earlier candidates for the action of AIDS [80].

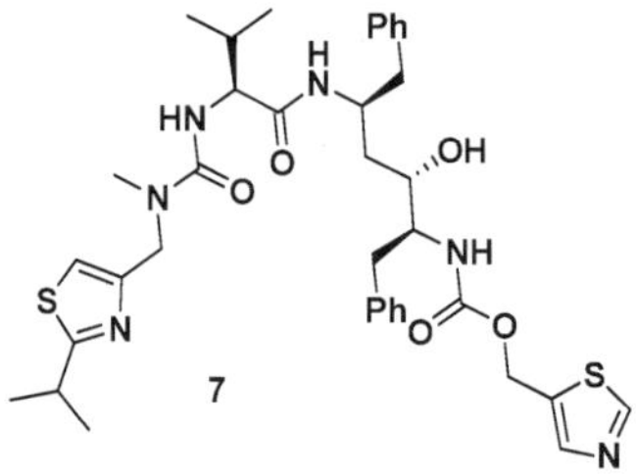

7

The dopamine D2-agonist pramipexole **27** (Mirapex) contains a fused bicyclic tetrahydrobenzothiazole design, which is also easy to obtain by a Hantzsch-type condensation reaction between a α-brominated protected form of 4-aminocyclohexanone and thiourea [81].

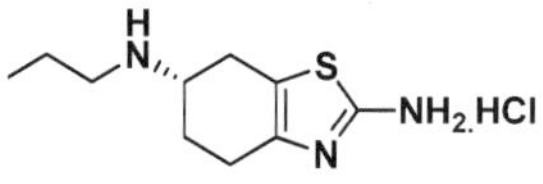

27

Famotidine (**28**, Pepcidine) is one of the top an H_2-receptor antagonists, which is equivalent to cimetidine that prevents various isoenzymes of the hepatic CYP450 system and the additional side effect (Swelling of the hands, feet or ankles) of enhancing the amount of gastric bacteria such as nitrate reducing bacteria. The arrangement of this ulcer therapeutic is very enthralling and contains a thiazole substituted guanidine and a sulfamoyl amidine. Current reports have performed designated famotidine as a significant ligand for numerous transition metals containing copper and cobalt developing tetradentate {*N, N,S,N*}-coordination spheres as revealed by single X-ray analysis [82]. Therefore, it seems viable that assured frequent bioavailable cations influence be included in the absorption and initiation of this thiazole involving compound. The formation of the thiazole ring [83,84] can be able again by condensation of thiourea with dichloroacetone.

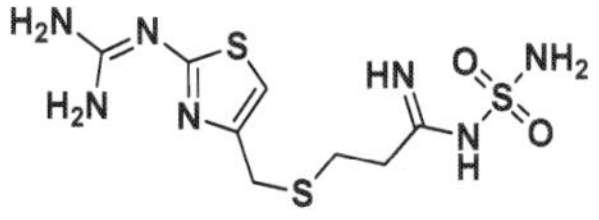

28

One more example of a thiazole ring enclosing drug is known in the unique xanthine oxidase inhibitor febuxostat **29** (Uloric) which was accepted by the FDA in 2009 [85]. This inhibitor works by hindering xanthine oxidase in a non-competitive manner. Subsequently, the quantity of the oxidation product uric acid is decreased. Thus, it is an extremely well-organized action for hyperuricemia in gout.

http://dx.doi.org/10.5772/62077

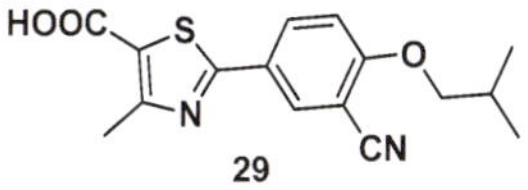

29

Takeuchi et al. described the total synthesis of the cyclic tripeptide histratamide H **30** established in the procedure of an extremely fluorous amino protecting group and multistep purifying by F-LPE using FC-72 in which 15 out of the 17 steps were purified by F-LPE [86].

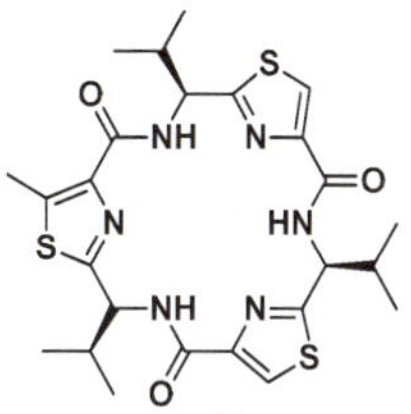

30

The construction of two heterocyclic rings in one synthetic step has been developed for the preparation of coumarin derivatives. In this process, the thiazole ring (**31 – 40**) is accomplished by Hantzsch reaction monitored by fabrication of pyrazole by reacting a 3-(2-bromoacetyl) coumarin with thiosemicarbazide and acetylacetone at room temperature [87].

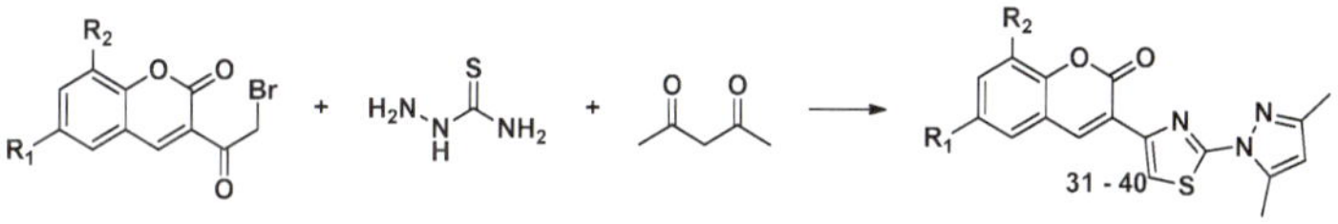

Adib et al. [88] described, in the latest work, a well-organized three component reaction that is significant to the formation of essential heterocycles titled by imidazo[1,2-*a*]thiazoles (**41** & **42**).

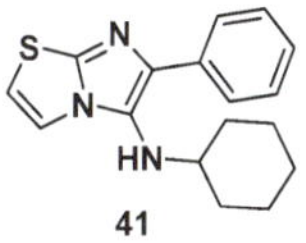

41

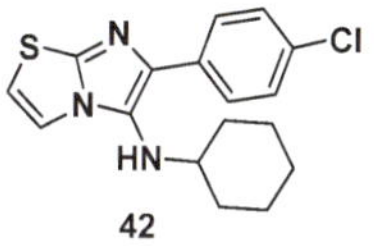

42

S. Zheng et. al. [89] synthesized five series of thiazole derivatives (**43 – 47**) for fascin therapeutic target as emerged from cancer cells is thoroughly related to tumor progression and metastasis. The entire compounds based on thiazole derivatives examined anti-migration and anti-invasion activities via possible inhibition of fascin function. The five series of analogs with elongated alkyl chain substitutions on the thiazole nitrogen revealed better anti-migration activities than those with other structural motifs.

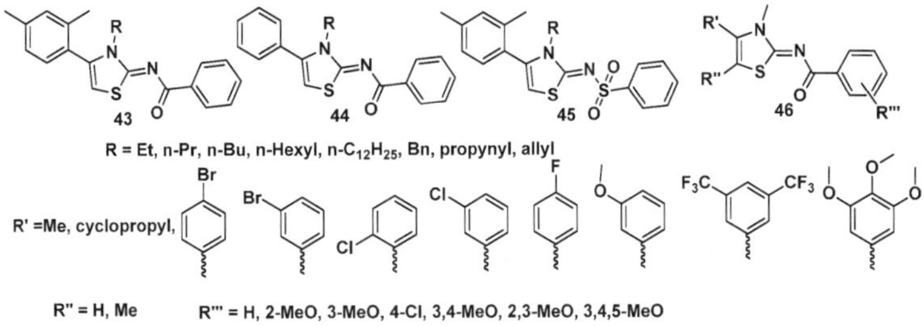

J. Zhu et.al. [90] reported that Human dihydroorotate dehydrogenase (HsDHODH) is a flavin-dependent mitochondrial enzyme that has been specialized as a prospective therapeutic aim for the medication of rheumatoid arthritis and other autoimmune diseases. On the basis of the main compound **48**, which was earlier recognized as potential HsDHODH inhibitor, a novel series of thiazole derivatives were designed and synthesized. The complex X-ray structures of the encouraging referents **49** and **50** established that these inhibitors bind at the recognized ubiquinone binding channel and directed us to explore additional potent inhibitors, such as compounds **44**, **46**, and **47** which exhibited double digit nanomolar activities of 26, 18, and 29 nM, respectively.

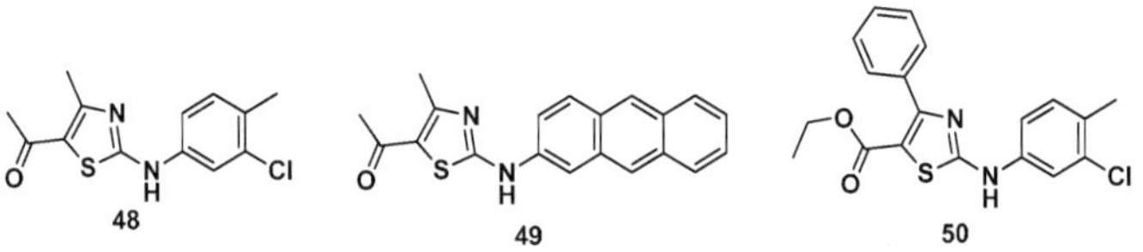

S. Singh et. al. [91] P-glycoprotein (P-gp) works as a therapeutic target for the improvement of multidrug conflict reversal agents. In this study, we synthesized twenty-one novel derivatives by peptide coupling at equivalent carboxyl and amino termini of (S)-valinebased bis-thiazole and mono thiazole derivatives with different chemical scaffolds. Consuming calcein-AM efflux assay, we recognized compound 51 (IC_{50} = 1.0 μM) containing 3,4,5-trimethoxybenzoyl and 2-aminobenzophenone groups, respectively, at the amino and carboxyl termini of the mono thiazole zwitterion. Compound 51 inhibited the photolabeling of P-gp with [^{125}I] - iodoarylazidoprazosin with IC_{50} = 0.75 μM and motivated the basal ATP hydrolysis of P-gp in a concentration-dependent manner (EC_{50} ATPase = 0.027 μM).

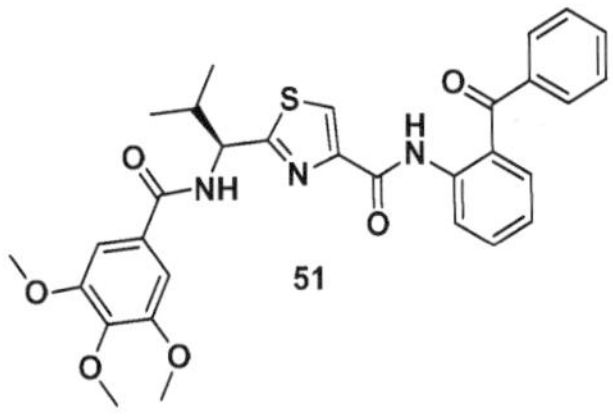

Oridonin **52**, a complex molecule *ent*-kaurane diterpenoid obtained from the traditional Chinese herb *Isodon rubescens*, has demonstrated great potential in the treatment of various human cancers due to its unique and safe anticancer pharmacological profile. However, with oridonin's poor solubility and poor bioavailability, hence C. Ding et. al.[92] inserted thiazole ring. The shortest way of synthesis of a series of novel nitrogen contained oridonin derivatives inserted thiazole-fused A-ring system through an active protecting group-free synthetic approach is the best of them, including compounds, **53–59** exhibited effective anti-proliferative effects against breast, pancreatic, and prostate cancer cells with low micromolar to submicromolar IC_{50} values as well as significantly improved aqueous solubility. These new derivatives achieved by realistically transforming the natural product have been established not only to induce considerably the apoptosis and inhibits the growth of triple-negative MDA-MB-231 breast cancer both in vitro and in vivo but also active against drug-resistant ER-positive MCF-7 clones.

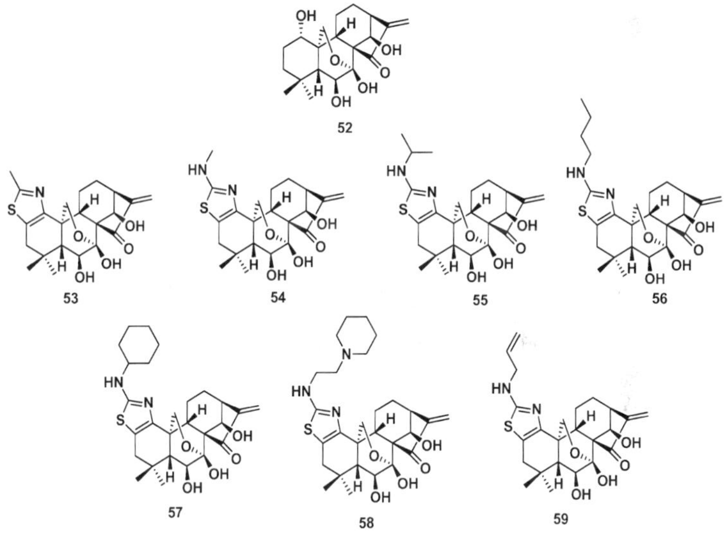

M. E. D. Francesco et. al. [93] reported a unique type of inhibitor, which designates the identification of a structurally various series of compounds including a 2-amino-1,3-thiazole as substitution of the carbamate in P4. Optimization studies motivated on structural variations in the P3, P2, and P1 regions of the macrocycle as well as on the linked chain caused the discovery of numerous analogs characterized by outstanding levels of enzyme and cellular activity. Among these, compound **60** exhibited the best pharmacokinetic profile in preclinical species and revealed constant liver levels subsequent oral administration in rats.

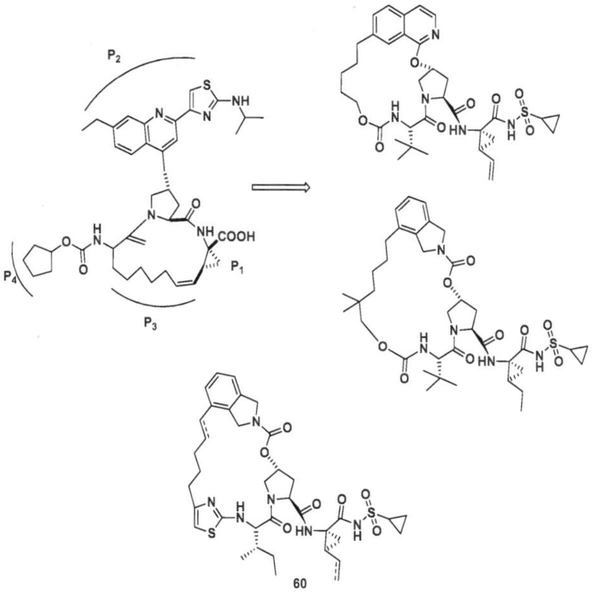

60

P. J. Sanfilippo et. al. [94] reported and described the synthesis and biological activity of a different kind of thiazole containing heterocycles as inhibitors of thrombin-induced human platelet aggregation. Additional estimation of selected compounds shows they inhibit platelet aggregation as motivated by a range of agonists. The highly active compounds also were established to inhibit fibrinogen binding to platelets. To further explain the mechanism of the action of these compounds, direct binding studies with the cleaned glycoprotein (GP) IIb/IIIa receptor were conducted. Flow cytometry analyzes of **61** and **62** designate that these compounds block the activation process of the GPIIb/IIIa receptor without denaturing the integrin receptor. On the basis of results, **62** showed the best profile as a novel non-peptide inhibitor of fibrinogen-mediated platelet aggregation.

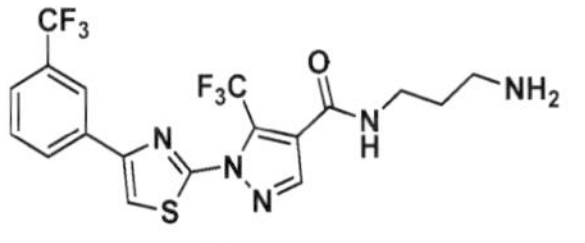

61

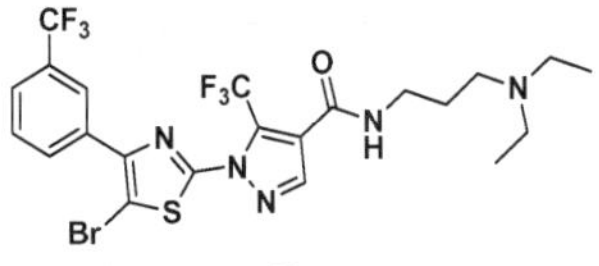

62

J. E. M. Koezen et. al. [95] prepared numerous *N*-[4-(2-pyridyl)thiazol-2-yl]benzamides, and these compounds exhibited adenosine affinities in the micromolar range. Most unexpected in the series of the *N*-[4-(2-pyridyl)thiazol-2-yl]amides were the retained adenosine affinities by the introduction of a cylopentanamide instead of the benzamide.

http://dx.doi.org/10.5772/62077

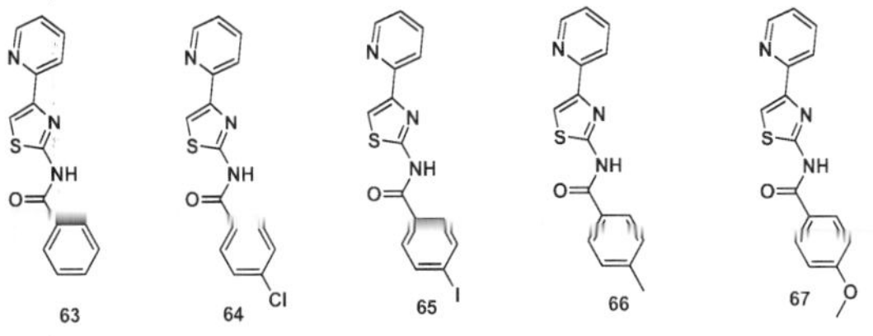

63 64 65 66 67

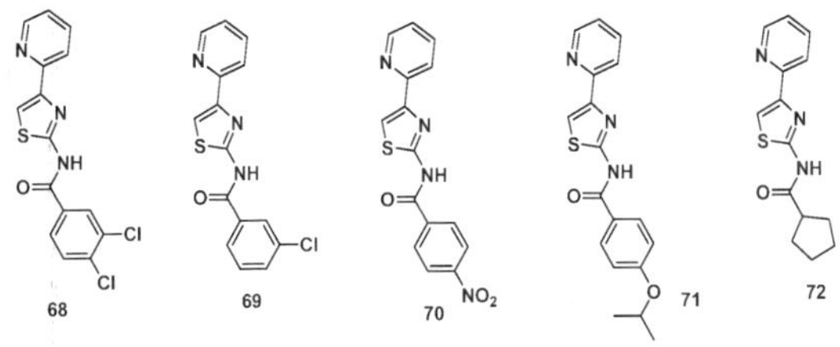

68 69 70 71 72

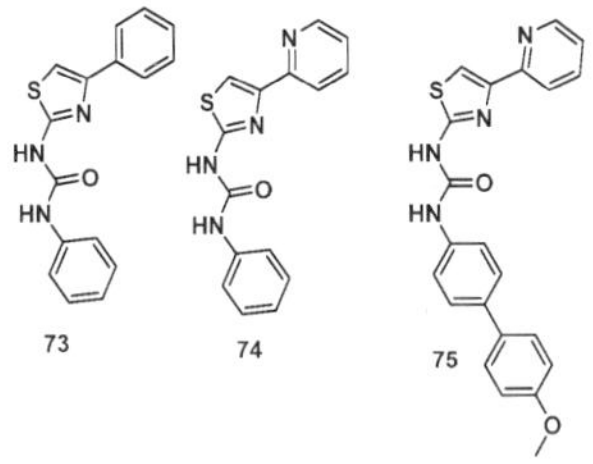

73 74 75

P. C. Srivastava et. al. [96] published a report in which they described the glycosylthiocarboxamides were used as the starting compounds for the synthesis of 2-D-ribofuranosylthiazole-4-carboxamide and 2-β-D-ribofuranosylthiazole-5-carboxamide (**76**). The structural variation of 2-β-D-ribofuranosylthiazole-4-carboxamide (**77**) into 2-(2,3,-5-tri-*O*-acetyl- β-D-ribofuranosyl)thiazole-4-carboxamide (**78**), 2-β-D-ribofuranosylthiazole-4-thiocarboxamide, and 2-(5-deoxy- β-D-ribofuranosyl)thiazole-4-carboxamide (**79**) is also designated. These thiazole nucleosides were verified for *in vitro* activity against type-1 herpes virus, type-3 parainfluenza virus, and type-13 rhinovirus and an *in vivo* test was run against parainfluenza virus. They were also analyzed as potential inhibitors of purine nucleotide biosynthesis. It was revealed that the compounds (**77** and **79**) which influenced the most noteworthy antiviral activity were also active inhibitors (40-70%) of guanine nucleotide biosynthesis.

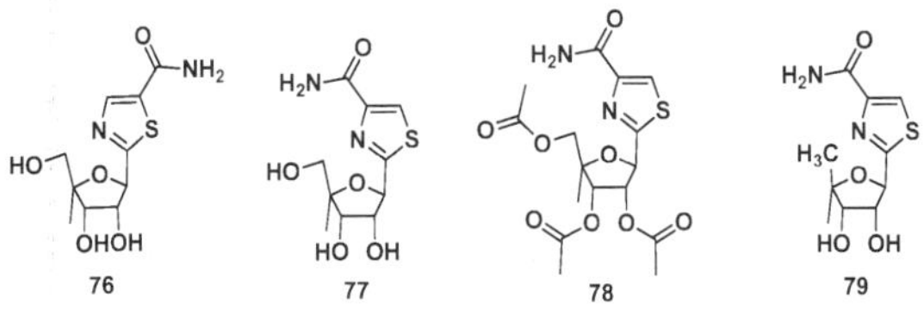

76 77 78 79

Z. Li et. al. [97] described the virtual screening data for flavivirus envelope proteins (E proteins) having been exposed to play a vital role in virus assembly, morphogenesis, and infection of host cells. Inhibition of flavivirus infection of a host cell by utilizing the small molecule envelope protein antagonist is an interesting approach to the development of antiviral agents. The virtual screening of the NCI Chemical database utilizing the dengue virus envelope protein structure showed numerous theoretical hit compounds. Bioassay consequences recognized a class of thiazole compounds with antiviral potency in cell-based analyzes. Variation of these lead compounds directed to a series of derivatives with enhanced antiviral activity and reduced cytotoxicity. The maximum activity exhibit compounds **80** and **81** were potent in the low micromolar concentration range in a cellular evaluate method.

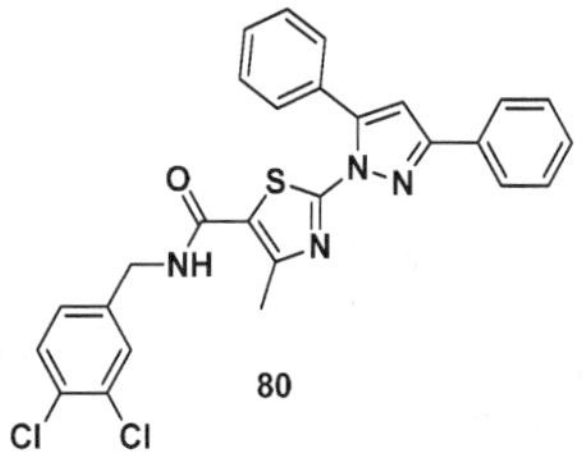

80

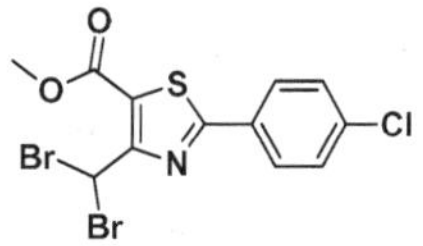

81

L. J. Lombardo et.al. [98] identified thiazole-based compounds as effective as Src/Abl kinase inhibitors with outstanding antiproliferative activity against hematological and solid tumor cell lines. Compound **82** was orally active in a K562 xenograft model of chronic myelogenous leukemia (CML), establishing complete tumor regressions and very low toxicity at multiple dose levels. On the basis of its powerful *in vivo* activity and promising pharmacokinetic profile, **82** was designated for supplementary characterization for oncology manifestations.

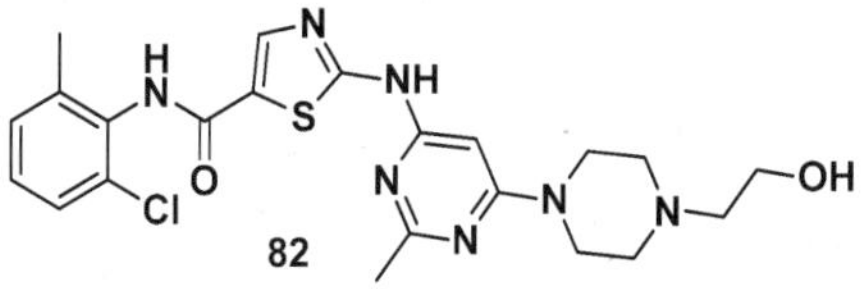

82

P. Madsen et. al. [99] explained the thiazole containing scaffold being potent human glucagon receptor antagonists with enhanced pharmacokinetic (PK) properties for expansion of pharmaceuticals for the medication of type-2 diabetes. The syntheses of compounds with cyclic moieties (5-aminothiazoles), their binding affinities for the human glucagon and GIP receptors, as well as affinities for mouse, pig, rat, dog, and monkey glucagon receptors. Normally, the compounds had less glucagon receptor affinity corresponding to compounds of the earlier series slightly, but this was rewarded for by much developed PK summaries in both rats and dogs with high oral bioavailabilities and constant high plasma coverages. The compounds exhibited species selectivity for glucagon receptor binding with very low affinities for the rat, mouse, rabbit, and pig receptors. However, dog and monkey glucagon receptor affinities seem to reflect the human situation. One of the compound sequence, **83**, was tested intravenously in an anesthetized glucagon-challenged monkey model of hyperglucagonaemia and hyperglycaemia and was revealed dose-dependently to reduce glycaemia

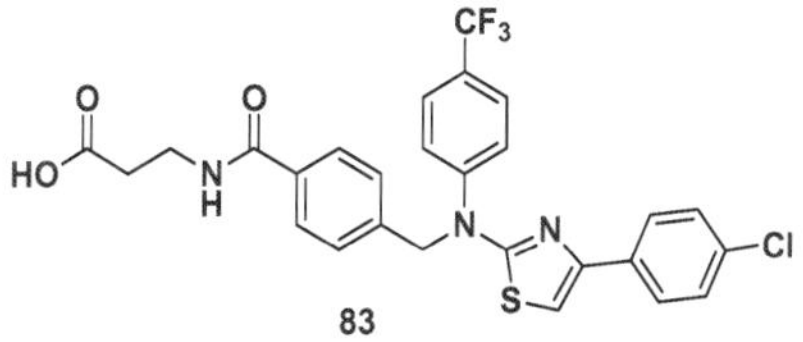

83

X. Cheng et.al. [100] reported a cell-based high throughput screening (HTS) operation for the search for potential candidates for octamer-binding transcription factor 4 (Oct3/4). In that process, they recognized numerous efficient small molecules for inducers of Oct3/4 expression. From HTS, optimized compounds are based on thiazole ring containing scaffold such as ethyl 2-([(4-chlorophenyl) amino])-thiazole-4-carboxylate, **84**, exhibiting high activity in implementing Oct3/4 expression. On the source of chemical expansion, once again screened the recognized derivatives requiring improved activities in the direction of Oct3/4 induction. Therefore, **84** and its analogs had afforded better potential small molecules proper for an iPSC generation.

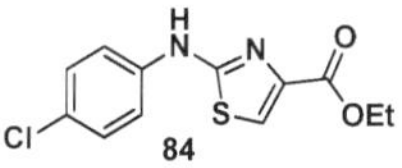

84

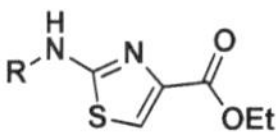

R = H (85), n-Bu (86), Allyl (87), Acetyl (88), -C_6H_5 (89), benzyl (90), ethyl 2-(phenylamino)thiazole-4-carboxylate (91), 3-Cl-C_6H_4 (92), 2-Cl-C_6H_4 (93), 4-Br-C_6H_4 (94), 3-Br-C_6H_4 (95), 2-Br-C_6H_4 (96), 4-F-C_6H_4 (97), 3-F-C_6H_4 (98), 2-F-C_6H_4 (99), 3,4-Cl_2-C_6H_3 (100), 2,4-Cl_2-C_6H_3 (101), 2,6-Cl_2-C_6H_3 (102), 3,6-Cl_2-C_6H_3 (103), 3,5-Cl_2-C_6H_3 (104)

C. P. Hencken et. al. [101] synthesized 23 new dehydroartemisinin (DART) trioxane analogs in which 11 thiazoles moiety-containing compounds remaining are based on two oxadiazoles, and ten carboxamides and screened them for in vitro activity in the Toxoplasma lytic cycle. Fifteen (65%) of the analogs were noncytotoxic to host cells ($TD_{50} \geq 320$ μM). Eight thiazole compounds exhibited effective inhibition of Toxoplasma growth (IC_{50} = 0.25-0.42 μM), similar in potency to artemether (IC50 = 0.31 μM) and >100 times stronger inhibitory than the presently working front-line drug trimethoprim (IC_{50} = 46 μM). The thiazoles as a ring were more efficient than other analogs at the inhibiting progress of extracellular as well as intracellular parasites. Surprisingly, two thiazole trioxanes (**109** and **110**) were parasiticidal; both inhibited parasite replication permanently after parasite contact to 10 μM of the drug for 24 h. However, the standard trioxane drugs artemisinin and artemether were not parasiticidal.

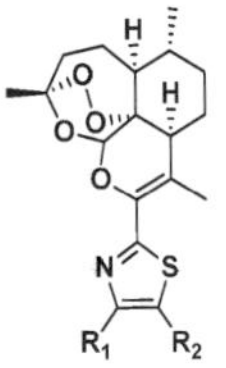

R_1, R_2 = H (105); R_1 = Me, R_2 = H (106); R_1, R_2 = Me (107); R_1 = Me, R_2 = H (108); R_1 = Ph, R_2 = H (109); R_1 = t-Bu, R_2 = H (110); R_1 = Et, R_2 = H (111); R_1 = C_6H_{11}, R_2 = H (112); R_1 = p-$CH_3C_6H_4$, R_2 = H (113); R_1 = p-$CH_3OC_6H_4$, R_2 = H (114); R_1 = p-$CH_3SC_6H_4$, R_2 = H (115); R_1 = p-$CH_3S(O)_2$Ph, R_2 = H (116).

Y. Kumar et. al. [102] reported that Methyl-4-(isothiocyanatomethy1)thiazole-2-carbamate have been obtained via chemical conversion containing 2-amino-4-(chloromethy1)thiazole (**117**) as precursor. The homoanalog, methyl 4-(2-isothiocyanatoethyl)thiazole-2-carbamate was synthesized via (2-aminothiazol-4-y1)acetic acid. All thiazole compounds synthesized were estimated for their capability to inhibit leukemia L1210 cell proliferation. Methyl 4-(isothiocyanatomethyl) thiazole-2-carbamate (**118**) was the active compound in this screen, inhibiting the growth of L1210 leukemic cells with an IC_{50} = 3.2 NM. Mitotic blocking performs to be its key mechanism of cytotoxic activity. Compound **118** furthermore was the only compound that confirmed important in uiua antifiiarial activity against the adult worms of Acanthocheilonema uiteae in experimentally infected jirds.

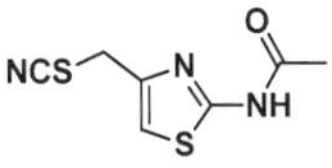

118

New thiazole based compounds [103] (1-(4-arylthiazol-2-yl)-2-(3-methylcyclohexylidene) - hydrazine) **119** are synthesized for the studied human B isoform of monoamine oxidase. These compounds were prepared as racemates and (R)-enantiomers by a stereoconservative synthetic arrangement in high yield and enantiomeric excess. The (S)-enantiomers of the highly active analogs have been separated by enantioselective HPLC. All compounds showed selective activity against hMAO-B with IC50 ranging between 21.90 and 0.018 μM.

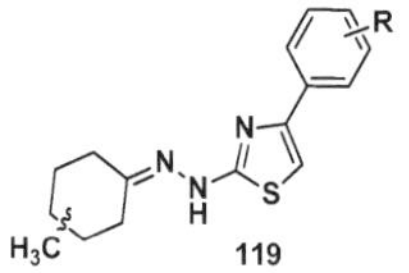

119

A. S. Mayhoub et.al. [104] synthesized a sequence of third-generation referents of methyl 4-(dibromomethyl)-2-(4-chlorophenyl)thiazole-5-carboxylate **120**, which had the highly potent antiviral activity comparable to the first and second generation derivatives, have been synthesized and verified against yellow fever virus consuming a cell-based assay. The compounds were aimed at the objectives of enlightening metabolic stability, therapeutic index, and antiviral potency. The biological effects of C4 and C5 substitution were studied. The methylthio ester and the dihydroxypropylamide analogs had the effective antiviral potencies and enhanced therapeutic indices and metabolic stabilities comparative to the parent compound **120**.

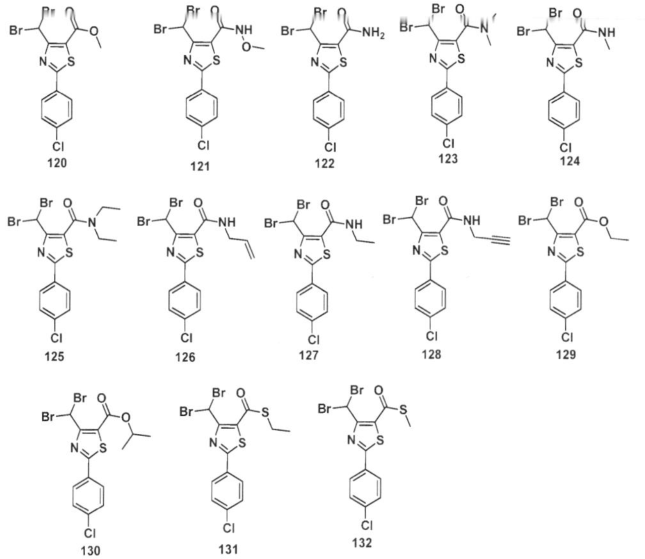

120 121 122 123 124 125 126 127 128 129 130 131 132

T. A. Dineen et.al. [105] reported the variation in structure **133** for the improved BACE1/CYP 3A4 inhibitors by a P1-phenyl ring of the hydroxyethylamine series to afford potent, which exhibit enhanced penetration into the CNS. Numerous compounds caused a robust decrease of Aβ levels in rat CSF and brain subsequently oral dosing, and compound **134** showed a better cardiovascular safety profile comparative to 133.

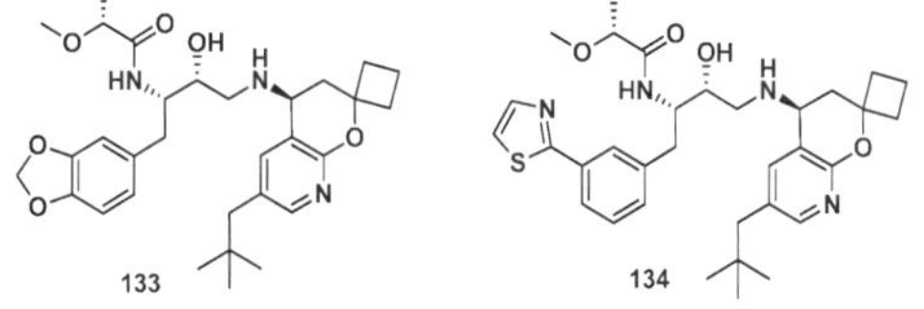

133 134

B. Ghosh et.al. [106] reported structure-activity relationship investigated on a unique hybrid sequence of derivatives where structural modification of aromatic hydrophobic moieties associated with the piperazine ring and bioisosteric exchange of the aromatic tetralin moieties were passed out. Binding assays were accepted with HEK-293 cells uttering either D2 or D3 receptors with tritiated spiperone to estimate inhibition constants (Ki). Functional activity of designated compounds in stimulating GTPγS binding was evaluated with CHO cells uttering human D2 receptors and AtT-20 cells uttering human D3 receptors. SAR results recognized compound 136 as one of the lead molecules with better agonist activity for D3 receptor (EC_{50} (GTPγS); D3= 0.52 nM; D2/D3 (EC_{50}): 223). Compounds **135** and **136** showed potent radical scavenging activity, the two lead compounds, **135** and **136**, showed more *in vivo* activity in two Parkinson's disease (PD) animal models, reserpinized rat model and 6-OHDA brought unilaterally lesioned rat model.

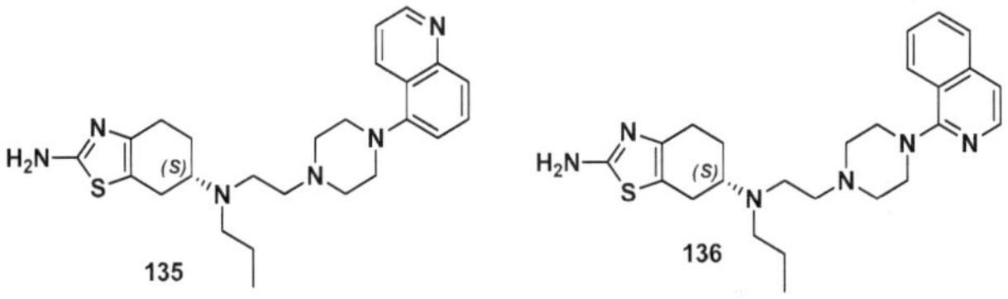

J. Das et.al. [107] explained that the 2-aminothiazole **137** was established as a unique Src family kinase inhibitor pattern through high calculated screening of their internal compound assembly. Optimization through consecutive structure-activity relationship iterations are recognized analogs **138** (Dasatinib, BMS-354825) and **139** as pan-Src inhibitors with nanomolar to subnanomolar strengths in cellular and biochemical assays. Molecular modeling techniques are utilized to conceptualize a recognized binding model for Lck inhibition by this type of compounds. The oral efficiency of this type of inhibitors was established with **139** in inhibiting the proinflammatory cytokine IL-2 *ex vivo* in mice (ED_{50} ~ 5 mg/kg) and in decreasing TNF levels in a serious murine model of inflammation. The oral efficiency of **139** was further verified in a chronic model of adjuvant arthritis in rats with recognized disease when ordered orally at 0.3 and 3 mg/kg two times daily.

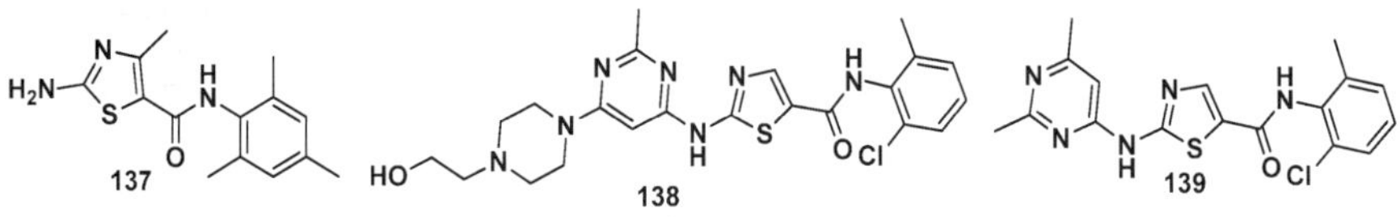

Major medicinal chemistry researcher focused on good docking small molecules inhibits the type 2 diabetes performances to have an insufficient or deficiency in one or both of these processes. Compounds that can activate glucokinase (GK) may serve as effective treatments for type 2 diabetes. In this process R. J. Hinklin et al. [108] reported that the recognition and preliminary optimization of a series of allosteric glucokinase activators (GKAs), revealed an early thiazolylamino pyridine-based hit that was elevated using a structure-based design approach and recognized **140** as an early lead. Compound **140** validated a good steadiness of *in vitro* effectiveness and enzyme kinetic limits and confirmed blood glucose decreases in oral glucose patience tests in both C57BL/6J mice and high-fat fed Zucker diabetic fatty rats.

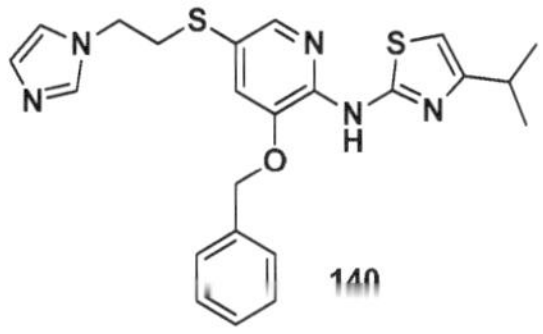

140

Spinal muscular atrophy (SMA), an inherited autosomal neurodegenerative disease, is the foremost genetic disorder disturbing infant mortality. Clinically, there are four kinds of SMA (types I, II, III, and IV). In fact, SMA is the top one genetic origin of death in children below the age of two, and several children life have been spoil due to confined to wheelchairs. There is presently no medication or effective treatment for SMA. Structure-activity relationships including microsomal stability, cell permeability, and *in vivo* pharmacokinetics (PK) studies are necessary. J. Xiao et al. [109] reported SMA active theoretically lead candidate selected from a sequence may work for as a valuable analysis for exploring the therapeutic aids of SMN protein up-regulation in SMA animal models and an initial point for clinical improvement. With regard to all the features including ADME properties, analogs **141** and **142** possessed the greatest combination of effectiveness, efficiency, mouse liver microsomal steadiness, and cell permeability of all the analogs that showed good activity.

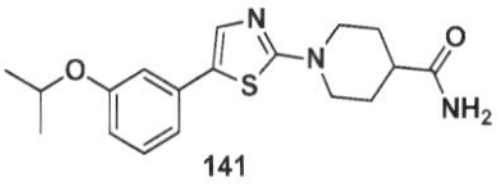

141

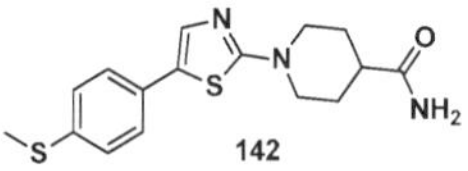

142

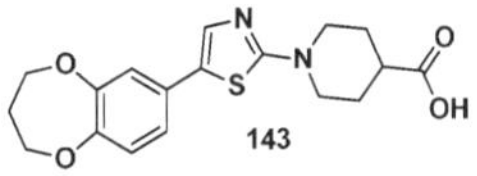

143

M. D, Rose, et al. [110] discussed the inhibition and antiviral activity consequence synthesis of 14- and 15-membered macrocycles for HIV-1 protease inhibitors (PIs) as obtained by ring-closing metathesis of the respective linear PIs. The macrocycles were very highly active than the linear precursors and compound 143, with a 2-thiazolyl ring was the best potent PI of this new series (K_i 2.2 nM, EC_{50} = 0.2 μM).

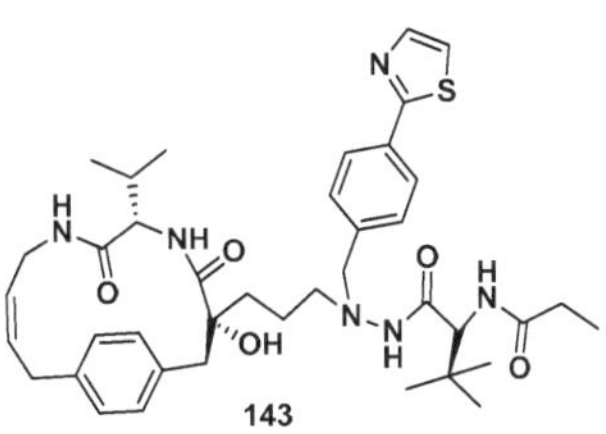

143

The preparation of a sequence of quinazolines inserted at C4 by aminothiazole ring is reported [111]. Their *in vitro* structure-activity relationships against Aurora A and B serine-threonine kinases are examined. The results reveal that quinazolines with a substituted aminothiazole at C4 possess potent Aurora A and B inhibitory activity and outstanding selectivity against a panel of several serine-threonine and tyrosine kinases. Compound **144** also found that the location and nature of the substituent on the thiazole play vital roles in cellular potency.

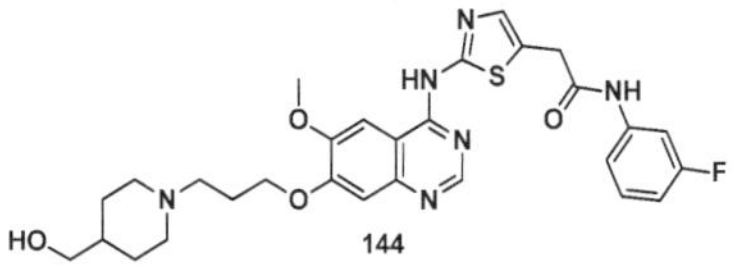

144

Approximately, the thiazole ring containing compound exhibits cathepsin K inhibitors [112]. The amalgamation of binding elements resulted at sub-250 pM, reversible, selective, and orally bioavailable cathepsin K inhibitors. In a series on of the compound exhibited single digit nanomolar inhibition *in vitro* (of rabbit osteoclastmediated degradation of bovine bone). The effective compound in this series, **145** (CRA-013783/ L-006235), was orally bioavailable in rats, with a terminal half-life of over 3 h, **145** was medicated orally in ovariectomized rhesus monkeys once per day for 7 days.

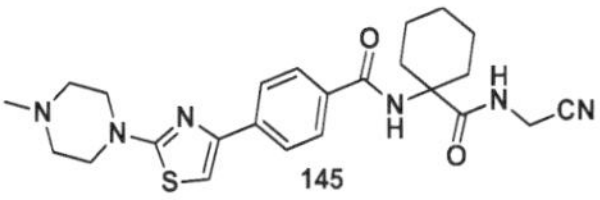

145

Haffner et. al. a series of thiazoloquinazolinones [113] were prepared and studied the inhibitory activity against CD38. Numerous compounds were also revealed to have good pharmacokinetic properties and established the capability to raise NAD levels in plasma, liver, and muscle tissue. Specifically, compound **146** was agreed to diet induced obese (DIO) C57Bl6 mice, enriching NAD > 5-fold in liver and >1.2-fold in muscle against control animals at a 2 h time point.

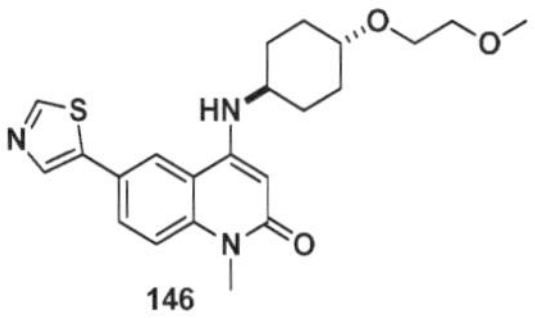

146

Thiazolo[5,4-d]pyrimidines and thiazolo[4,5-d]pyrimidines are structurally mimic with purines, in which a 1,3-thiazole ring system exchanges the imidazole moiety. While purine chemistry is broadly discussed in the literature, the number of medicinal chemistry publications that reported the synthesis and biological studies of thiazolopyrimidines is narrow comparable with purines. Seemingly, the thiazolopyrimidine scaffold is not very often used in drug discovery platforms. However, biological activities of unequivocal thiazolo[4,5-d]pyrimidines and thiazolo[5,4-d]pyrimidines have been described. A summary of available compounds with their biological significance is presented in Figures 147, **148**, **149** and **150**.

http://dx.doi.org/10.5772/62077

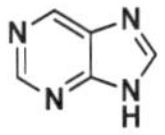

9*H*-purine

147

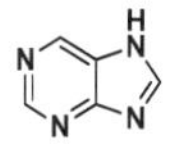

7*H*-purine

148

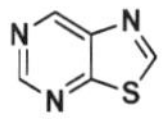

thiazolo[5,4-*d*]pyrimidine

149

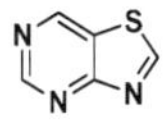

thiazolo[4,5-*d*]pyrimidine

150

Thiazolo[4,5-d]pyrimidine derivative **151** revealed *in vivo* activity towards a broad range of RNA and DNA viruses [114] and also had antitumor and antimetastatic activity [115]. The guanine analogs **152** exhibited potent in vitro activity against human cytomegalovirus (HCMV) [116]. Thiazolo[4,5-d]pyrimidine-5,7-dione analogs (compound **153**) have been described as having potential anti-inflammatory activities, because of TNF inhibition [117]. 4 2-Oxo-3-arylthiazolo[4,5-d]pyrimidine analogs (compound **154**) have been produced as antagonists of the corticotrophin-releasing hormone (CRH) R1 receptor [118]. 2-Thio-3-aryl-thiazolo[4,5-d]pyrimidine and its derivatives have been reported as having anticancer (compound **155**) [119], antimicrobial and anti-inflammatory activity (compound **156a & 156b**) [120]. 2-Aminothiazolo[4,5-d]pyrimidines (compound **157a & 157b**) which performance as CXCR2 receptor antagonists are also recognized [121]. Lately, 2,7-substituted-thiazolo[4,5-d]pyrimidines (compound **158**) have been explained as ATP-competitive inhibitors of protein kinase [122].

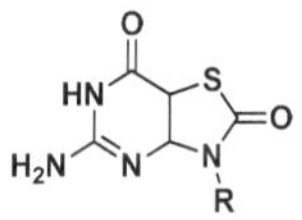

R = β-D-ribofuranosyl (151)
R = n-pentyl (152)

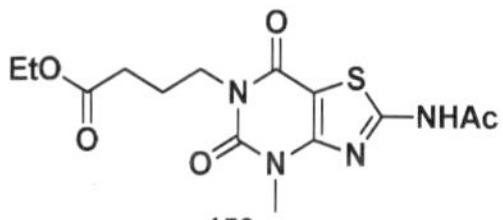

153

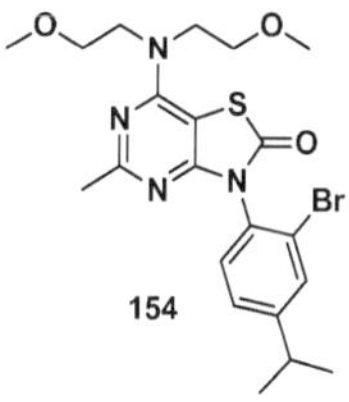

154

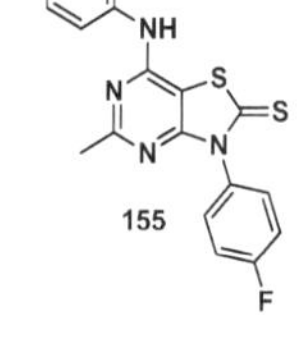

155

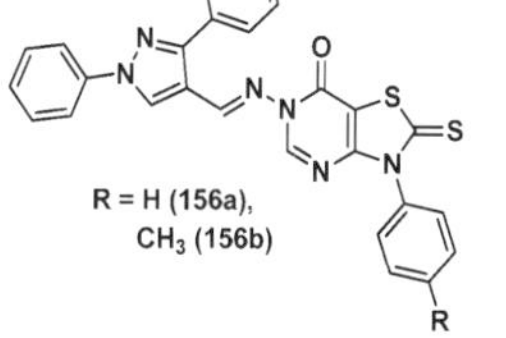

R = H (156a), CH$_3$ (156b)

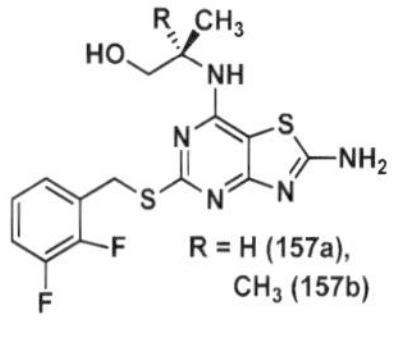

R = H (157a), CH$_3$ (157b)

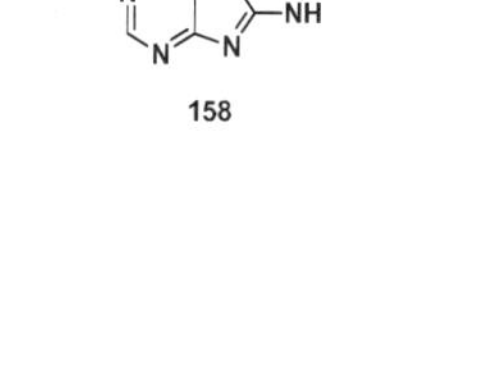

158

2,5-Diaminothiazolo[5,4-d]pyrimidin-7(6H)-one (Compound **159**), a thio-isostere of 8-aminoguanine, was established to be a poor inhibitor of purine nucleoside phosphorylase (PNP) [123]. 7-Diethylamino-5-methylthiazolo[5,4-d]pyrimidine **160** has vasodilating and hypotensive properties, inhibits platelet aggregation, and decreasing cholesterol levels [124]. Thiazolo[5,4-d]pyrimidines were enclosed by numerous patent properties such as activators of caspases and inducers of apoptosis (compound **161**) [125], anti-angiogenic agents (compound **162**) [126], growth factor receptor inhibitors (compound **163**) [127], heat shock protein 90 (HSP-90) inhibitors (compound **164**) [128], and xanthine oxidase inhibitors (compound **165**) [129].

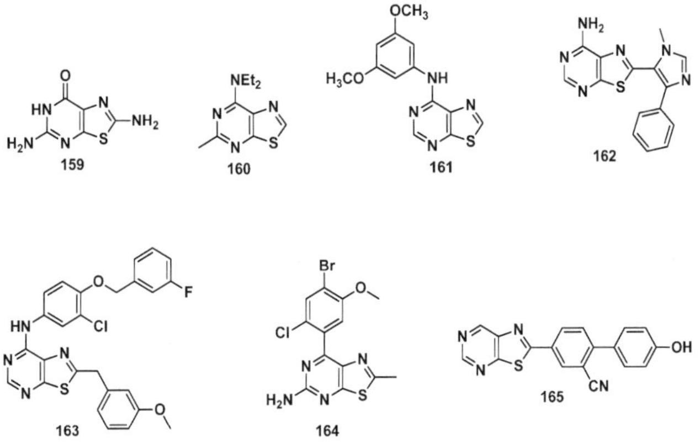

5. General Synthetic Routes to Thiazolo[5,4-d]pyrimidines

In wide-ranging, pyrimidines with a nitrogen-containing substituent at position 5 (such as an amino or nitro group) can work as precursors for the formation of thiazolo[5,4-d]pyrimidines by thiazole ring condensation. 5-Amino- or 5-nitropyrimidines can be organized from diethyl amino-, nitro-, or acetylamino-malonate by reacts with coupling reagents such as thiourea [130], urea [131], guanidine [132] and amidines [133] in alkali conditions. By reaction of the 4,6-dihydroxypyrimidine analog with a thionation reagent (Lawesson's reagent or phosphorus pentasulfide) in pyridine, alteration of oxygen into sulfur and thiazole ring closure is accomplished. Interaction of 5-amino-6- mercaptopyrimidines with reagents such as phosgene [134], formic acid [135], and acid anhydride [136] also gives thiazolo[5,4-d]pyrimidines.

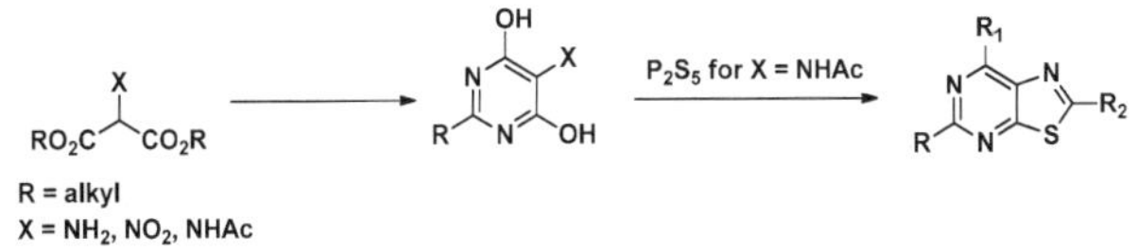

http://dx.doi.org/10.5772/62077

Thiazolo[5,4-d]pyrimidine-1-N-oxides, ready to obtain from 6-chloro-1,3-dimethyl-5-nitro-pyrimidinone by reaction with mercapto compounds, monitored by base catalyzed dehydrative cyclization, can be simply deoxygenated to produce thiazolopyrimidines. Reductive deoxygenation by treatment of the thiazolopyrimidine oxides with sodium dithionite or oxidative deoxygenation with dimethylformamide at reflux temperature can produce the anticipated thiazolopyrimidines [137].

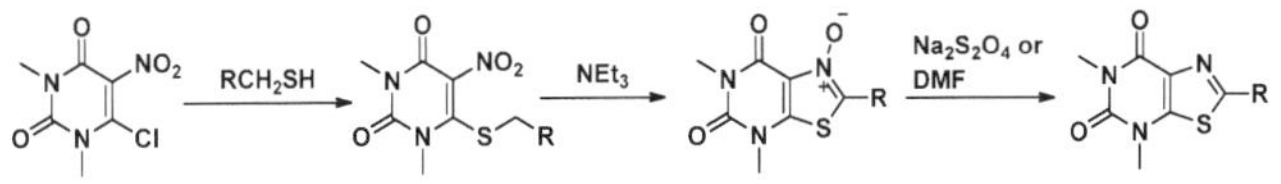

2-Mercaptothiazolo[5,4-d]pyrimidines easily obtained from 6-chloro-5-nitropyrimidines by the reaction with carbon disulfide and sodium sulfide [138].

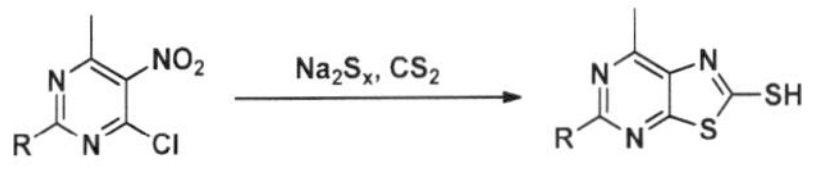

2-Amino-7-chlorothiazolo[5,4-d]pyrimidines are prepared from 5-amino-4,6-dichloropyrimidine and isothiocyanate in presence base [139].

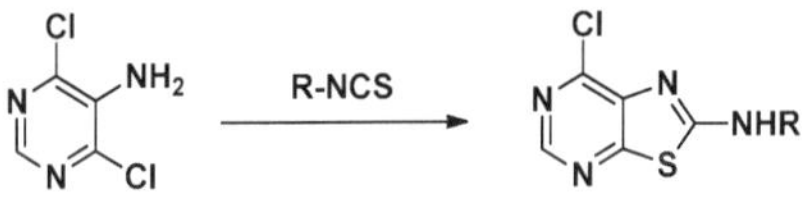

Ahmed et al. [140] reported the synthesis of thiazolo[5,4-d]pyrimidines from pyrimidines without 5-amino or 5-nitro substituents. The reaction between 5-bromo-4-thioxo-pyrimidinones and dimethylcyanamide affords carbodiimide intermediates, which is a very fast intramolecular cyclization to produce a thiazole ring.

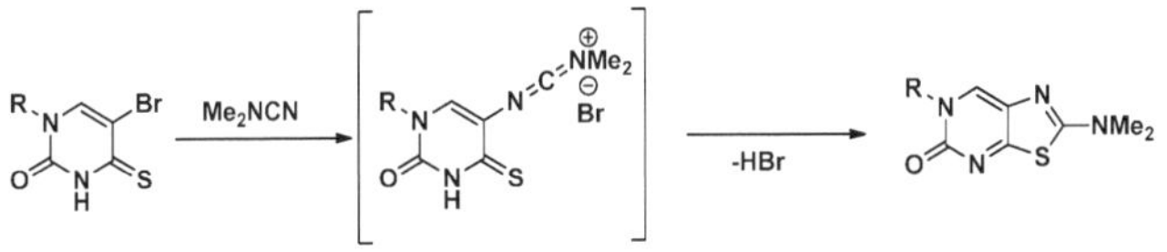

On the other hand, thiazolo[5,4-d]pyrimidines also obtained from 5-aminothiazole derivatives, are prepared from aminomalononitrile (or its derivatives) and isothiocyanates [141] or thioesters [142]. The next to 5-amino and 4-cyano (or conforming carboxamide or ester groups) on the thiazole ring are proper functionalities to concept a fused pyrimidine ring system. 7-Aminothiazolo[5,4-d]pyrimidines can be prepared from 5-amino-4-cyanothiazoles by reaction with reagents such as orthoesters and amidines [143,144]. The reaction between 5-amino-4-carboxamide (or carboxylate) thiazoles and orthoesters [145], formamide [146], and ethyl chloroformate/DMF [147] gives thiazolo[5,4-d]pyrimidin-7(6H)ones.

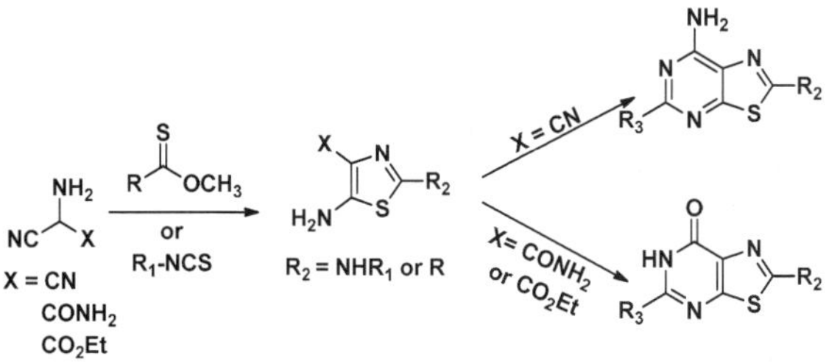

Additionally, Thiazolo[5,4-d]pyrimidinones readily to obtained from the corresponding oxazolopyrimidines. In fact, that 1,3-oxazole ring system is quickly converted into a 1,3-thiazole by a thermal rearrangement. The thioamide replaced oxazole derivative is prepared from the corresponding amide by reacts with Lawesson's reagent. Heating generates the nitrile ylide by electrocyclic ring opening, followed by a 1,5-dipolar electrocyclization affording the thiazole [148].

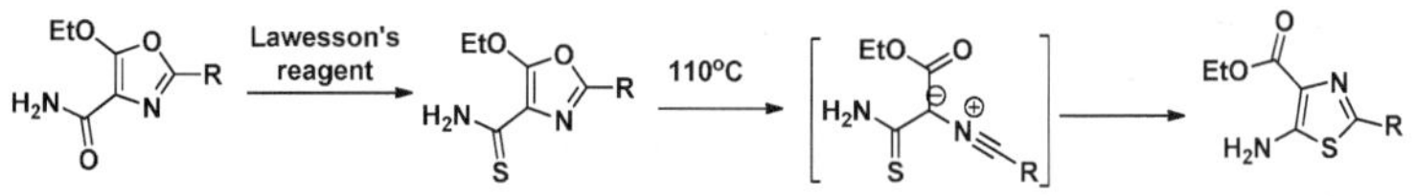

6. General Synthetic Routes to Thiazolo[4,5-d]pyrimidines

The preparation of thiazolo[4,5-d]pyrimidines from a properly substituted pyrimidine compound yields 2-aminothiazolo[4,5-d]pyrimidines. Thiocyanation of 6- aminopyrimidines reacts with potassium thiocyanate, bromine and pyridine, proceeded by cyclization yields the 2-aminothiazolo[4,5-d]pyrimidines [149].

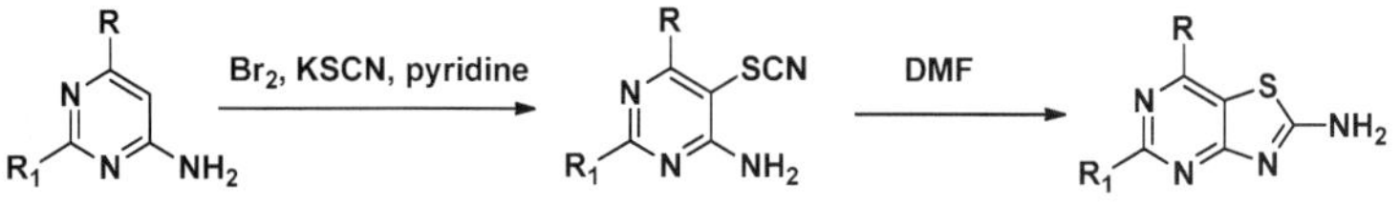

Condensation of 5-bromobarbituric acid with thiourea and/or its derivatives in the presence of an alkali yields thiazolo[4,5-d]pyrimidine derivatives [150].

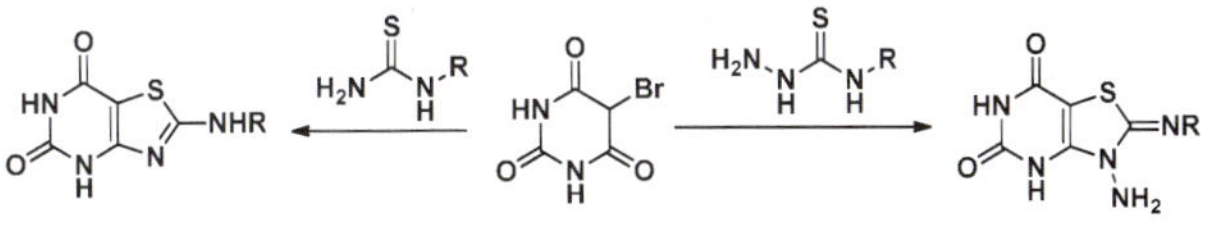

Another significant technique for the preparation of thiazolo[4,5-d]pyrimidines from thiazoles is via the aza-Wittig reaction [151]. The iminophosphorane intermediates are found from 4-chloro-5-formylthiazoles which reacts with sodium azide and triphenylphosphine (Staudinger reaction). In Addition, reaction with isocyanates affords the corresponding carbodiimides, followed by heating, and undergo an electrocyclic ring closing, which upon a Dimroth-type rearrangement obtained thiazolo[4,5-d]pyrimidines [152].

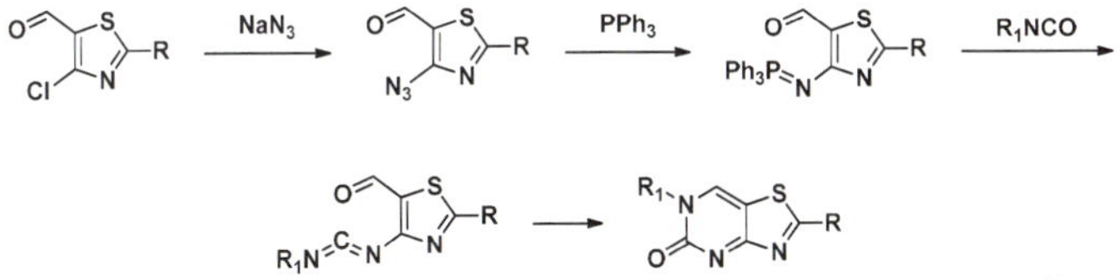

It is needed to expand the hit ratio in HTS campaigns; fortunate molecular scaffold systems offer a perfect basis of main compounds. A particular library created on preferable bioisosteres groups are inserted into the main scaffold and can generate the bioactive compounds in a broad range of biological tests. Numerous researchers have developed these structures in such a fashion. For example, Ghorpade and co-workers built a library based on the thiazolopyridines privileged scaffold [153] whereas Bebernitz and co-workers made use of the chlorosulfonic acid combined with thiazolopyridines scaffold [154] (compounds **166** and **167**)

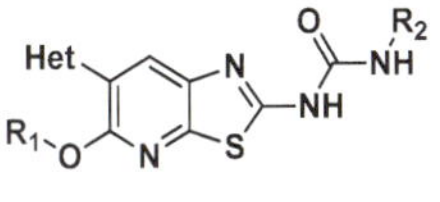

166

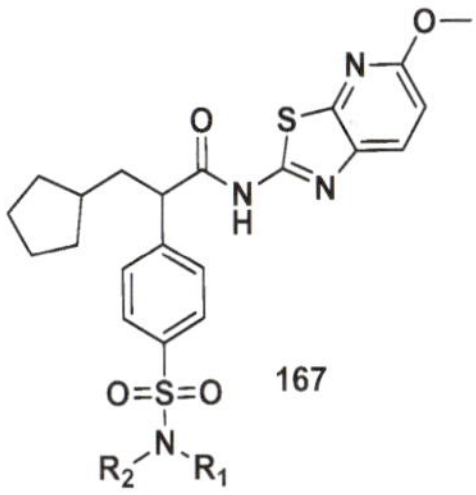

167

Bicyclic nitrogen, sulfur - containing heterocycles, such as Thiazolo[5,4-b]quinoline, thiazolo-pyridines, and thiazolopyrimidines are well-known pharmacophores in drug discovery [155-157]. Examples of promoted drugs with a bicyclic core structure include AMG-369 analogs performing as Lysophospholipid edg1 (S1P1) and Receptor Agonists Lysophospholipid edg8 (S1P5) Receptor Agonists [158]. Thiazole sulfonamides based scaffold, used as antidepressants and for the treatment of Vasopressin (AVP) V1b Antagonists [159]. Kirsch and co-workers described a solution-phase synthesis of 7-amino-thiazolo[4,5-b]pyridine derivatives [160] as well as fused-pyridine analogs such as the thiopheno[2,3-b]pyridines [161] using the Friedlander reaction.

Thiazolo[4,5-b]pyridine derivatives reveal a broad range of biological properties. For example, thiazolo[4,5-b]pyridines have confirmed actions as serine protease factor Xa (fXa) inhibitors for thrombosis [162], as metabotropic glutamate receptor 5 (mGluR5) antagonists for several CNS syndromes [163], as histamine H3-receptor antagonists for epilepsy and Alzheimer's disease [164], as epidermal growth factor receptor (EGFR) tyrosine kinase inhibitors [165], and as cAMP phosphodiesterase (PDE) III inhibitors for congestive heart catastrophe [166].

T. Lee et. al. [167] reported a series of 2,5,6,7-tetrasubstituted thiazolo[4,5-b]pyridine derivatives (**168**) from solid-phase synthesis. Thorpe-Ziegler type cyclization of solid supported cyanocarbonimidodithioate with α-halo ketones gave thiazole resin, which were transformed to the preferred thiazolopyridine resin by the Friedlander procedure under microwave irradiation conditions. After oxidation of sulfides to sulfones, nucleophilic desulfonative substitution with amines yielded the target thiazolo[4,5-b]pyridine derivatives.

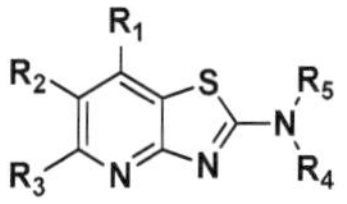

168a - 168ax

Y. Takahashi et. al [168] described the synthesis and structure–activity relationships of a unique series of 7-dialkylamino-3-phenyl-6-methoxy pyrazolo[5,1-b]thiazole derivatives to utilize as selective antagonists of the corticotropin-releasing factor 1 (CRF1) receptor. The best favorable compound, N-butyl-3-[4-(ethoxymethyl)-2,6-dimethoxyphenyl]-6-methoxy-N-(tetrahydro-2H-pyran-4-yl)pyrazolo-[5,1-b, 1,3]thiazole-7-amine (**169**), exhibited very high affinity (IC_{50} = 70 nM) and functional antagonism (IC_{50} = 7.1 nM) for the human CRF1 receptor.

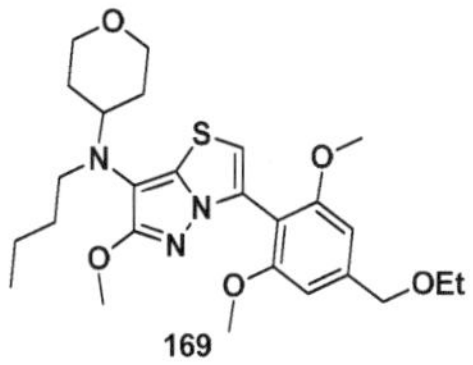

169

7. Summary

This chapter discusses the high synthetic perspective of several methods for synthesis of thiazoles and its derivatives that have been published in the last three decades. Many pharmaceutically active heterocycles have been obtained based on the reaction of acid hydrazides particularly concerning Hantzsch reaction, Dimroth type rearrangement, Tchernich reaction,

CooK–Heilbron reaction, Gabriel reaction, Erlenmeyer reaction, Hartke–Seib reaction and Dubs reaction. Fundamentally α-halo carbonyl compounds and substituted thiourea or thiosemicarbazide are potential precursors for the creation of wide range of thiazole analogous as main synthon constituents for generation of several diverse heterocycles. The aza-Wittig product such as iminophosphorane intermediates obtained from 4-chloro-5-formylthiazoles by treatment with sodium azide and triphenylphosphine (Staudinger reaction) with most other various reagents like isocyanate, isothiocyanate and carbondisulfide for bicyclic generation system containing thiazole moiety under basic, acidic or neutral reaction conditions. Most of these reagents are available from simply or commercially accessible, inexpensive precursors. This chapter has also verified the noticeable feature to the advancement of an eco-friendly experimental technique for the synthesis of heterocyclic compounds. The synthetic approaches showed in this chapter can be comprehensive to the synthesis of natural macrocyclic thiazole ring containing heterocycles and also suggest that α-halo carbonyl compounds can be a favorable building block in combinatorial synthesis of functionalized heterocyclic derivatives used for the design of unique very active pharmaceutical drugs with a broad spectrum of bioresponses. In certain cases, reports on the less yield of bioactive heterocycles in this chapter could be overwhelmed by forthcoming synthetic chemists with this sustained research and new methods for extensive approach and explained experimental procedures could be explored for its development for generation of a library of such multi-functional heterocycles to afford a useful encouragement to medicinal chemistry.

Acknowledgements

I acknowledge the Department of Chemistry, Nizam College, Osmania University. Moreover, grateful to Dr. Ravi Varala, Head, IIIT, Basara, India, for the valuable support to write this chapter.

Author details

Someshwar Pola*

Address all correspondence to: somesh.pola@gmail.com

Department of Chemistry, Nizam College, Osmania University, Hyderabad, Telangana, India

References

[1] Drews J, Drug Discovery: A Historical Perspective, Science; 2000. P. 1960 - 1964. DOI: 10.1126/ science. 287.5460.1960

[2] Wess G, Urmann M, Sickenberger B, Medicinal Chemistry: Challenges and Opportunities; Angew. Chem. Int. Ed.; 2001. p. 3341 - 3350. DOI: 10.1002/1521-3773(20010917)40:18<3341::AID-ANIE3341>3.0.CO;2-D

[3] Lombardino J. G. Lowe J. A, A guide to drug discovery: The role of the medicinal chemist in drug discovery — then and now, Nat. Rev. Drug Discov.: 2004. p. 853 – 862. DOI: 10.1038/nrd1523

[4] Kolb H.C, Finn M. G, Sharpless K. B, Click Chemistry: Diverse Chemical Function from a Few Good Reactions, Angew. Chem. Int. Ed.: 2001. p. 2004 - 2021. DOI: 10.1002/1521-3773(20010601)40:11<2004::AID-ANIE2004>3.0.CO;2-5

[5] Horton D. A, Bourne G. T, Smythe M. L, The Combinatorial Synthesis of Bicyclic Privileged Structures or Privileged Substructures, Chem. Rev.: 2003. p. 893 -930. DOI: 10.1021/cr020033s

[6] Zoltewicz J. A, Deady L. W, "Quaternization of Heteroaromatic Compounds: Quantitative Aspects", Advances in Heterocyclic Chemistry, Elsevier: 1978. p. 71–121. DOI: 10.1016/S0065-2725(08)60103-8

[7] Sammes P. G, Comprehensive Organic Chemistry" Heterocyclic compounds, Part 20-1, Ed., Pegammon Press, Oxford: Vol. 4, 1979. p. 976

[8] Elederfield R.C, "Heterocyclic compounds", (Five membered Heterocycles containing two Hetero-atoms), Ed., John Wiley and Sons Inc., New York: 1961.

[9] Breslow R, On the Mechanism of Thiamine Action. IV.1 Evidence from Studies on Model Systems, J. Am. Chem. Soc.: 1958. p. 3719 - 3726. DOI: 10.1021/ja01547a064

[10] Borisenko V. E, Koll A, Kolmakov E. E, Rjasnyi A. G, Hydrogen bonds of 2-aminothiazoles in intermolecular complexes (1:1 and 1:2) with proton acceptors in solutions, J. Mol. Struct.: 2006. p. 101 – 115. DOI:10.1016/j.molstruc.2005.08.006

[11] Bhargava P.N, Lakhan R, Tripathi K, J. Ind. Chem. Soc., 1982, 59, 773; (b) Maulard T, Lagorce J. F, Thomos J. C, Raby C, Biological Evaluation of Compounds with -NCS-Group or Derived from Thiazole and Imidazole. Activity on Prostaglandin Synthetase Complex, J. Pharm. Pharmacol.: 1993. p. 731 - 735. DOI: 10.1111/j.2042-7158.1993.tb07098.x

[12] Maj J, Rog Z, Skuza G, Kolodziejezyk K, Antidepressant effects of pramipexole, a novel dopamine receptor agonist, J. Neural. Transm.: 1997. p. 525 - 533.

[13] Milne G. W. A, Handbook of Antineoplastic Agents, Gower; Ashgate, Ed., London, UK, 2000.

[14] M. V. N. De Souza, M. V. De Almeida, Drugs anti-HIV: past, present and future perspectives, Quim. Nova.: 2003. p. 366 – 372. DOI.org/10.1590/S0100-40422003000300014

[15] Markus B, Salome von G, Christoph B, Werner J P, Molecular Aspects of Drug Recognition by Specific T Cells, Curr. Drug. Tar.: 2004. p. 1 – 11. DOI: 10.2174/1389450033347082

[16] Knadler M P, Bergstrom R. F, Callaghan J. T, Rubin A, Nizatidine, an H2-blocker. Its metabolism and disposition in man, Drug Metab. Dispos.: 1986. p. 175 - 182.

[17] Lednicer D, Mitscher L. A, George G. I, Organic Chemistry of Drug Synthesis, Wiley, New York, Vol. 4, 1990, p. 95.

[18] Rehman M. Z, Anwar C. J, Ahmad S, An Efficient Synthesis of 2-Alkyl-4-hydroxy-2H-1,2-benzothiazine-3-carboxamide-1,1-dioxides, Bull. Korean Chem. Soc.: 2005. p. 1771 - 1775. DOI: http://dx.doi.org/10.5012/ bkcs.2005.26.11.1771

[19] Beuchet P, Varache-Lembege M, Neveu A, Leger J. M, Vercauteren J, Larrouture S, Deffieux G, Nuhrich A, New 2-sulfonamidothiazoles substituted at C-4: synthesis of polyoxygenated aryl derivatives and in vitro evaluation of antifungal activity, Eur. J. Med. Chem.: 1999. p. 773 - 779. DOI:10.1016/S0223-5234(99)00215-9

[20] Fink B. E, Mortensen D. S, Stauffer S. R, Aron Z. D, Katzenellenbogen J. A. Novel structural templates for estrogen-receptor ligands and prospects for combinatorial synthesis of estrogens, Chem. Biol.: 1999. p. 205 - 219. DOI: 10.1016/ S1074-5521(99)80037-4

[21] van Muijlwijk-Koezen J. E, Timmerman H, Vollinga R. C, von DrabbeKunzel J. F, de Groote M, Visser S, IJzerman A. P. Thiazole and Thiadiazole Analogues as a Novel Class of Adenosine Receptor Antagonists, J. Med. Chem.: 2001. p. 749 - 762. DOI: 10.1021/jm0003945

[22] Füßlein, Martin, Bretschneider, Thomas, Fischer, Reiner, Jeschke, Peter, Köhler, Adeline, Kluth, Joachim, Mühlthau, Friedrich August, Voerste, Arnd, Sato, Yoshitaka, United States Patent: 20110319428, 2011, December.

[23] Wang Q, Li H, Li Y, Huang R. Synthesis and Herbicidal Activity of 2-Cyano-3-(2-chlorothiazol-5-yl)methylaminoacrylates,J. Agric. Food Chem.: 2004. p. 1918 - 1922. DOI: 10.1021/jf035312a

[24] Tintcheva I, Maximova V, Deligeorgiev T, Zaneva D, Ivanov I. New asymmetric monomethine cyanine dyes for nucleic-acid labelling: absorption and fluorescence spectral characteristics, J. Photochem. & Photobiol. A: Chem.: 2000. p. 7 - 11. DOI: 10.1016/S1010-6030(99)00207-5

[25] Rucker V. C, Foister S, Melander C, Dervan P. B. Sequence Specific Fluorescence Detection of Double Strand DNA, J. Am. Chem. Soc.: 2003. p. 1195 - 1202. DOI: 10.1021/ ja021011q

[26] Shibamoto T. Heterocyclic compounds found in cooked meats, J. Agric. Food Chem.: 1980. p. 237 - 243. DOI: 10.1021/jf60228a031

[27] Buchanan J. L, Bohacek R. S, Luke G. P, Hatada M, Lu X, Dalgarno D.C, Narula S. S, Yuan R, Holt D. A. Structure-based design and synthesis of a novel class of Src SH2 inhibitors, Bioorg. Med. Chem. Lett.: 1999. p. 2353 - 2358. DOI: 10.1016/S0960-894X(99)00388-1

[28] Hallinan E. A, Hagen T. J, Tsymbalov S, Stapelfeld A, Savage M. A. 2,4-Disubstituted oxazoles and thiazoles as latent pharmacophores for diacylhydrazine of SC-51089, a potent PGE2 antagonist, Bioorg. Med. Chem.: 2001. p. 1 - 6. DOI: 10.1016/S0968-0896(00)00229-7

[29] Das B, Reddy S. V, Ramu R. A rapid and high-yielding synthesis of thiazoles and aminothiazoles using ammonium-12-molybdophosphate, J. Mol. Catal. A: Chem.: 2006. p. 235 - 237. DOI: 10.1016/ j.molcata.2006.02.065

[30] Narender M, Reddy S. M, Sridhar R, Nageswar Y, V. D, Rao R. K.. Aqueous phase synthesis of thiazoles and aminothiazoles in the presence of β-cyclodextrin, Tetrahedron Lett.: 2005. p. 5953 -5955. DOI: 10.1016/j.tetlet.2005.06.130

[31] Siddiqui H. L, Iqbal A, Ahmed S, Weaver G. Synthesis and Spectroscopic Studies of New Schiff Bases, Molecules: 2006. p. 206 - 211. DOI: 10.3390/11020206

[32] Karade H, Sathe M, Kaushik M.P., An efficient method for the synthesis of 2-aminothiazoles using silica chloride as a heterogeneous catalyst, Catal. Commun.: 2007. p. 741 - 746. DOI: 10.1016/j.catcom.2006.09.005

[33] Egido E. G, Wong S. Y. F, Warrington B. H., A Hantzsch synthesis of 2-aminothiazoles performed in a heated microreactor system, Lab Chip, 2002. p.31 - 33. DOI: 10.1039/B109360F

[34] George W. K, Arjun R. M. Microwave promoted synthesis of functionalized 2-aminothiazoles, Tetrahedron Lett.: 2006. p. 5171 - 5172. DOI: 10.1016/j.tetlet.2006.05.053

[35] Metzger J. V. "Thiazole and its derivatives (Part I)", An Interscience Publication: 1979. p. 165 - 335. DOI: 10.1002/9780470187081

[36] Willstatter R, Wirth T. Über Thioformamid, Berichte: 1909. p. 1908 - 1922. DOI: 10.1002/cber.19090420267

[37] Otto H, Process for the production of Thiazoles, U. S. Patent No. 2160867: 1939.

[38] George Y. S, Subhi A. A. Preparation and spectral characterization of substituted 2-aminothiazoles, J. Chem. Eng. Data: 1973. p. 99 - 102. DOI: 10.1021/je60056a029

[39] G. Marchesini, Gazz. Chim. Ital., 1893, 23, 437.

[40] G. Marchesini, Gazz. Chim. Ital., 1894, 24, 65.

[41] Clarke H T, Gurin S, Studies of Crystalline Vitamin B1. XII. The Sulfur-Containing Moiety, J. Am. Chem. Soc.: 1935. p. 1876 - 1881. DOI: 10.1021/ja01313a035

[42] Todd A R, Bergel F, Karimullah. Über Aneurin, II. Mitteil.: Über die Synthese von N-Arylthiazoliumsalzen; über Einzelheiten in der Konstitution des Aneurins und Thiochroms, Berichte: 1936. p. 217 - 223. DOI: 10.1002/cber.19360690139

[43] Karimullah, Substituted phenyl- and benzyl-thiazolium salts, J. Chem. Soc.: 1937. p. 961 - 962. DOI: 10.1039/JR9370000961

[44] Kiprianov A. I, Asnina F. I, Shako I. K.. J. Gen. Chem., U.S.S.R., 1948, 18, 165.

[45] E. D. Sych, E. K. Mikitenko, Khim. Geterotsikl. Soedinenii, 1971, 7, 857.

[46] George W. W. Reactions in the thiazole series. ii. the reaction of 2-chlorobenzothiazole with thiourea in aqueous media, J. Org. Chem.: 1939. p. 436 - 441. DOI: 10.1021/jo01216a010

[47] Ian H. S. Concerning amino-acids, peptides, and purines, J. Chem. Soc.: 1949. p. 2099 - 2107. DOI: 10.1039/JR9490002099

[48] Cook A. H, Heilbron I, MacDonald S. F, Mahadevan A. P. Studies in the azole series. Part XII. Some thiazolopyrimidines, J. Chem. Soc.: 1949. p. 1064 - 1068. DOI: 10.1039/JR9490001064

[49] Cook A. H, Heilbron I, Mahadevan A. P. Studies in the azole series. Part XI. The interaction of α-amino-nitriles, hydrogen sulphide, and ketones, J. Chem. Soc.: 1949. p. 1061 – 1064. DOI: 10.1039/JR9490001061

[50] Cook A. H, Heilbron I. M, Levy A. L. Studies in the azole series. Part II. The interaction of α-amino-nitriles and carbon disulphide, J. Chem. Soc.: 1947. p. 1598 - 1609. DOI: 10.1039/JR9470001598

[51] Cook A. H, Heilbron I. M, Stern E. Studies in the azole series. Part X. Some 5-amino-2-mercapto-4-alkylthiazoles and 2 : 4-dithio-5-alkylhydantions, J. Chem. Soc.: 1948. p. 2031 - 2033. DOI: 10.1039/JR9480002031

[52] Hoggarth E. The rearrangement of 2-benzenesulphenamidothiazoles. Part II. Thiazole compounds substituted in position 4, J. Chem. Soc.: 1947. p.114 - 118. DOI: 10.1039/JR9470000114

[53] Gabriel S. Synthese von Oxazolen und Thiazolen II, Berichte: 1910. p. 1283 - 1287. DOI: 10.1002/cber.19100430219

[54] Asinger F, Thiel M, Pallas E. Über die gemeinsame Einwirkung von elementarem Schwefel und gasförmigem Ammoniak auf Ketone I. Die Gemeinsame Einwirkung von Schwefel und Ammoniak auf Diäthylketon, Annalen: 1957. p. 37 – 49. DOI: 10.1002/jlac.19576020104

[55] Davies W, MacLaren J, Wilkinson L. The synthesis and properties of 2 : 4-diaminothiazoles J. Chem. Soc.: 1950. p. 3491 - 3494. DOI: 10.1039/JR9500003491

[56] Chase B. H, Walker J. A novel reaction of thiourea, the structure of Jaffé's base, and related studies, J. Chem. Soc.: 1955. p. 4443 - 4450. DOI: 10.1039/JR9550004443

[57] Alessandro D, Alberto M. Thiazole-Mediated Synthetic Methodology, Chem. Rev.: 2004. p. 2557-2599. DOI: 10.1021/cr0200791

[58] Koichi N, Shen S. S, Mao-Yi L. Intramolecular nucleophilic substitution at an sp^2 carbon: synthesis of substituted thiazoles and imidazole-2-thiones, Tetrahedron Lett.: 2009. p. 3161 - 3163. 10.1016/j.tetlet.2008.12.118

[59] Dominique R, Qiao Q, Goodnow Jr. R. 2,4-Disubstituted-5-acetoxythiazoles: useful intermediates for the synthesis of thiazolones and 2,4,5-trisubstituted thiazoles, Tetrahedron Lett.: 2008. p. 3682 - 3686. 10.1016/j.tetlet.2008.03.140

[60] Sharma R N, Xavier F P, Vasu K K, Chaturvedi S C, Pancholi S S. Synthesis of 4-Benzyl-1,3-thiazole derivatives as potential anti-inflammatory agents: An Analogue-based Drug Design Approach, J Enz Inhib Med Chem.: 2009. p. 890 – 897. DOI: 10.1080/14756360802519558

[61] Clark R F, Zhang T, Wang X, Wang R, Zhang X, Camp H S, Beutel B A, Sham H L, Gu Y J. Phenoxy thiazole derivatives as potent and selective acetyl Co-A carboxylase-2 inhibitors: Modulation of isozyme selectivity by incorporation of phenyl ring substituent, Bioorg Med Chem Lett.: 2007. p.1961-1965. DOI: 10.1016/j.bmcl.2007.01.022

[62] Mahajan M. P, Sodnhi S. M, Ralhan N. K. Studies on Heterocyclics. V. Synthesis of Thiazolotriazepines, Bull. Chem. Soc. Jpn.: 1976. p. 2651 - 2652. DOI: 10.1246/bcsj.49.2651

[63] Astrakhantseva N. I, Zhiryakov V. G, Abramenko P. I. Synthesis of 2-mercapto-substituted thieno[2,3-d]thiazoles and ben zothieno[3,2-d] thiazoles, Chem. Het. Compd.: 1976. p. 1123 - 1124.

[64] Beyer H, Lassing W, Ruhlig G. Über Thiazole, XV. Mitteil.: Über die Kondensation von Chloraceton mit Thiosemicarbazid und einige Umlagerungen in der Thiazolreihe, Chem. Ber.: 1953. p. 764 - 769. DOI: 10.1002/cber.19530860621

[65] Haviv F, Ratajczyk J. D, DeNet R. W, Kerdesky F. A, Walters R. L, Schmidt S. P, Holms J. H, Young P. R, Carter G. W, 3-[1-(2-Benzoxazolyl)hydrazino]propanenitrile derivatives: inhibitors of immune complex induced inflammation, J. Med. Chem.: 1988. p.1719 - 1728. DOI: 10.1021/jm00117a010

[66] Patt W. C, Hamilton H. W, Taylor M. D, Ryan M. J, Jr. Taylor D. G, Connolly C. J. C, Structure-activity relationships of a series of 2-amino-4-thiazole-containing renin inhibitors, J. Med. Chem.: 1992. p. 2562 -2572. DOI: 10.1021/jm00092a006

[67] Bell F. W, Cantrell A. S, Hoegberg M, Jaskunas S. R, Johansson N. G, Jordon C. L, Kinnick M. D, Lind P, Jr. Morin J. M, Noreen R, Oberg B, Palkowitz J. A, Parrish C. A. Phenethylthiazolethiourea (PETT) Compounds, a New Class of HIV-1 Reverse

Transcriptase Inhibitors. 1. Synthesis and Basic Structure-Activity Relationship Studies of PETT Analogs, J. Med. Chem.: 1995. p. 4929 - 4936. DOI: 10.1021/jm00025a010

[68] Fink B. A, Mortensen D. S, Stauffer S. R., Aron Z. D, Katzenellenbogen J. A. Novel structural templates for estrogen-receptor ligands and prospects for combinatorial synthesis of estrogens, Chem. Biol.: 1999. p. 205 -219. DOI: 10.1016/S1074-5521(99)80037-4

[69] Van Muijlwijk-Koezen J. E, Timmerman H, Vollinga R. C, Von Drabbe Kunzel J. F, De Groote M, Visser S, Ijzerman A. P. Thiazole and Thiadiazole Analogues as a Novel Class of Adenosine Receptor Antagonists, J. Med. Chem.: 2001. p. 749 - 762. DOI: 10.1021/jm0003945

[70] Metzger J. V. "Comprehensive Heterocyclic Chemistry Part-I", Pergamon: New York, 1984. p. 328.

[71] Sherif A. F. R, Ibrahim M. El-A, Heba A. A. Design and synthesis of some thiazolyl and thiadiazolyl derivatives of antipyrine as potential non-acidic anti-inflammatory, analgesic and antimicrobial agents, Bioorg. Med. Chem.: 2009. p.882 – 895. DOI: 10.1016/j.bmc.2008.11.035

[72] Zablotskaya A, Segal I, Germane S, Shestakova I, Domracheva I, Nesterova A, Geronikaki A, Lukevics E. Silyl Modification of Biologically Active Compounds. 8. Trimethylsilyl Ethers of Hydroxyl-containing Thiazole Derivatives, Chem. Het. Comp.: 2002. p. 859 – 866.

[73] Dae-Kee K, Joon H. C, Young J. A. Synthesis and biological evaluation of 5-(pyridin-2-yl)thiazoles as transforming growth factor-β type1 receptor kinase inhibitors, Bioorg. Med. Chem. Lett., 2008. p.2122 - 2127. DOI: 10.1016/j.bmcl.2008.01.084

[74] Rajan S, Hardik G, Thaker M, Tony G, Kamala K. V. Design, synthesis and characterization of novel 2-(2,4-disubstituted-thiazole-5-yl)-3-aryl-3H-quinazoline-4-one derivatives as inhibitors of NF-κB and AP-1 mediated transcription activation and as potential anti-inflammatory agents, Eur. J. Med. Chem.: 2009. p. 2184 - 2189. DOI: 10.1016/j.ejmech.2008.10.031

[75] Johan D.O, et. al. Discovery of a potent and selective Aurora kinase inhibitor, Bioorg. Med. Chem. Lett.: 2008. p. 4880 - 4884. DOI: 10.1016/j.bmcl.2008.07.073

[76] El-Subbagh H. I, Al-Obaid A. M. 2,4-Disubstituted thiazoles II. A novel class of antitumor agents, synthesis and biological evaluation, Eur. J. Med. Chem.: 1996. p.1017 - 1021. DOI: 10.1016/S0223-5234(97)86181-8

[77] Guay D. R. P. Cefdinir: An advanced-generation, broad-spectrum oral cephalosporin, Clin. Ther.: 2002. p. 473 – 489. DOI: 10.1016/S0149-2918(02)85125-6; (b) Mooney M. T, Deguchi S, Tada T, Fujioka M, Okamoto Y, Yasuda T. Interaction of Cefdinir with Iron in Aqueous Solution, Chem. Pharm. Bull.: 1995. p. 374 – 377. DOI: org/10.1248/cpb.43.374; (c) González M, Rodriguez Z, Tolón B, Rodriguez J. C, Veléz H, Valdez B,

Lopéz M. A, Fini A. An alternative procedure for preparation of cefdinir, Farmaco: 2003. p. 409 - 418. DOI: 10.1016/S0014-827X(03)00063-6

[78] Kempf D. J, Marsh K. C, Denissen J. F, McDonald E, Vasavanonda S, Flentge C. A, Green B. E, Fino L, Park C. H, Kong X. P. ABT-538 is a potent inhibitor of human immunodeficiency virus protease and has high oral bioavailability in humans, Proc. Natl. Acad. Sci. U. S. A.: 1995. p. 2484 - 2488. DOI:10.1073/pnas.92.7.2484

[79] Kerdesky F. A. J, Seif L. S. A Novel and Efficient Synthesis of 5-(Hydroxymethyl) Thiazole: An Important Synthon for Preparing Biologically Active Compounds, Synth. Commun.: 1995. p. 2639 - 2645. DOI: 10.1080/00397919508011810

[80] Schneider C, Griss G, Hurnaus R, Kobinger W, Pichler L, Bauer R, Mierau J, Hinzen D, Schingnitz G, Eur. Patent 186087B1, 1986.

[81] Juarez J. M, Silva Guisasola L. O. Method for the resolution of 2-amino-6-propylamino-4,5,6,7-tetrahydrobenzothiazol and intermediate compounds, WO2006120268: 2006.

[82] Miodragović D. U, Bogdanović G. A, Miodragović Z. M, Radulović M. D, Novaković S. B, Kaluđerović G. N, Kozłowski H. Interesting coordination abilities of antiulcer drug famotidine and antimicrobial activity of drug and its cobalt(III) complex, J. Inorg. Biochem.: 2006. p. 1568 - 1574. DOI: 10.1016/j.jinorgbio.2006.05.009

[83] Sprague J. M, Land A. H, Ziegler C, Derivatives of 2-Amino-4-methylthiazole, J. Am. Chem. Soc.: 1946. p. 2155 - 2159. DOI: 10.1021/ja01215a010

[84] Hirata Y, Yanagisawa I, Ishii Y, Tsukamoto S, Ito N, Isamura Y, Takeda M, U.S.: Patent 4283408, 1981.

[85] Hasegawa M. A Facile One-Pot Synthesis of 4-Alkoxy-1,3-benzenedicarbonitrile, Heterocycles: 1998. p. 857 - 864. DOI: 10.3987/COM-97-S(N)89

[86] Nakamura Y, Takeuchi S, Synthesis of Bistratamide H Using a New Fluorous Protecting Group, QSAR & Combinatorial Science: 2006. p. 703 – 708. DOI: 10.1002/qsar.200640053; (b) Nakamura Y, Okumura K, Kojima M, Takeuchi S. An expeditious synthesis of bistratamide H using a new fluorous protecting group, Tetrahedron Lett.: 2006. p. 239 - 243. DOI: 10.1016/j.tetlet.2005.10.144

[87] Rao V. R, Sumanth K. A facile one step synthesis of 3-[2-(3,5-dimethyl-1H-pyrazol-1-yl)-4-thiazolyl]-2H-1-benzopyran-2-ones under solvent free conditions, J. Chem. Res.: 2002. p. 420 - 421. DOI: org/10.3184/ 030823402103172725

[88] Adib M, Mahdavi M, Noghani M. A, Mirzaei P. Catalyst-free three-component reaction between 2-aminopyridines (or 2-aminothiazoles), aldehydes, and isocyanides in water, Tetrahedron Lett.: 2007. p. 7263 -7265. DOI: 10.1016/j.tetlet.2007.08.049

[89] Zheng S, Zhong Q, Xi Y, Mottamal M, Zhang Q, Schroeder R. L, Sridhar J, He L, McFerrin H, Wang G, Modification and Biological Evaluation of Thiazole Derivatives

as Novel Inhibitors of Metastatic Cancer Cell Migration and Invasion, J. Med. Chem.: 2014. p.6653 - 6667. DOI: 10.1021/jm500724x

[90] Zhu J, Han L, Diao Y, Ren X, Xu M, Xu L, Li S, Li Q, Dong D, Huang J, Liu X, Zhao Z, Wang R, Zhu L, Xu Y, Qian X, Li H. Design, Synthesis, X-ray Crystallographic Analysis, and Biological Evaluation of Thiazole Derivatives as Potent and Selective Inhibitors of Human Dihydroorotate Dehydrogenase, J. Med. Chem.: 2015. p. 1123 - 1139. DOI: 10.1021/jm501127s

[91] Singh S, Prasad N. R, Chufan E. E, Patel B. A, Wang Y. J, Chen Z. S, Ambudkar S. V, Talele T. T, Design and Synthesis of Human ABCB1 (P-Glycoprotein) Inhibitors by Peptide Coupling of Diverse Chemical Scaffolds on Carboxyl and Amino Termini of (S)-Valine-Derived Thiazole Amino Acid, J. Med. Chem.: 2014. p. 4058 – 4072. DOI: 10.1021/jm401966m; (b) Singh S, Prasad N. R, Kapoor K, Chufan E. E, Patel B. A, Ambudkar S. V. Talele T. T. Design, Synthesis, and Biological Evaluation of (S)-Valine Thiazole-Derived Cyclic and Noncyclic Peptidomimetic Oligomers as Modulators of Human P-Glycoprotein (ABCB1), ChemBioChem: 2014. p. 157 - 169. DOI: 10.1002/cbic.201300565

[92] Ding C, Zhang Y, Chen H, Yang Z, Wild C, Chu L, Liu H, Shen Q, Zhou J. Novel Nitrogen-Enriched Oridonin Analogues with Thiazole-Fused A-Ring: Protecting Group-Free Synthesis, Enhanced Anticancer Profile, and Improved Aqueous Solubility, J. Med. Chem.: 2013. p. 5048 - 5058. DOI: 10.1021/jm400367n

[93] Francesco M. E. D, Dessole G, Nizi E, Pace P, Koch U, Fiore F, Pesci S, Muzio J. D, Monteagudo E, Rowley M, Summa V. Novel Macrocyclic Inhibitors of Hepatitis C NS3/4A Protease Featuring a 2-Amino-1,3-thiazole as a P4 Carbamate Replacement, J. Med. Chem.: 2009. p. 7014 - 7028. DOI: 10.1021/jm900524b

[94] Sanfilippo P. J, Urbanski M. J, Beers K. N, Eckardt A, Falotico R, Ginsberg M. H, Offord S, Press J. B, Tighe J, Tomko K, Gordon P. A. Novel Thiazole-Based Heterocycles as Selective Inhibitors of Fibrinogen-Mediated Platelet Aggregation, J. Med. Chem.: 1995. p. 34 – 41. DOI: 10.1021/jm00001a008

[95] van Muijlwijk-Koezen J. E, Timmerman H, Vollinga R. C, Frijtag von Drabbe Kunzel J, de Groote M, Visser S, IJzerman A. P. Thiazole and Thiadiazole Analogues as a Novel Class of Adenosine Receptor Antagonists, J. Med. Chem.: 2001. p. 749 - 762. DOI:10.1021/jm0003945

[96] Srivastava P. C, Pickering M. V, Allen L. B, Streeter D. G, Campbell M. T, Witkowski J. T, Sidwell R. W, Robins R. K. Synthesis and antiviral activity of certain thiazole C-nucleosides, J. Med. Chem.: 1977. p.256 - 262. DOI: 10.1021/jm00212a014

[97] Li Z, Khaliq M, Zhou Z, Post C. B, Kuhn R. J. Cushman M. Design, Synthesis, and Biological Evaluation of Antiviral Agents Targeting Flavivirus Envelope Proteins, J. Med. Chem.: 2008. p. 4660 – 4671. DOI: 10.1021/jm800412d; (b) Suh J, Yum E. K, Cheon H. G, Cho Y. S. Synthesis and Biological Evaluation of N-aryl-4-aryl-1,3-Thia-

zole-2-Amine Derivatives as Direct 5-Lipoxygenase Inhibitors, Chemical Biology & Drug Design: 2012. p.89 - 98. DOI: 10.1111/j.1747-0285.2012.01371.x

[98] Lombardo L. J, Lee F. Y, Chen P, Norris D, Barrish J. C, Behnia K, Castaneda S, Cornelius L. A. M, Das J, Doweyko A. M, Fairchild C, Hunt J. T, Inigo I, Johnston K, Kamath A, Kan D, Klei H, Marathe P, Pang S, Peterson R, Pitt S, Schieven G. L, Schmidt R. J, Tokarski J, Wen M. L, Wityak J, Borzilleri R. M. Discovery of N-(2-Chloro-6-methyl- phenyl)-2-(6-(4-(2-hydroxyethyl)- piperazin-1-yl)-2-methylpyrimidin-4- ylamino)thiazole-5-carboxamide (BMS-354825), a Dual Src/Abl Kinase Inhibitor with Potent Antitumor Activity in Preclinical Assays, J. Med. Chem.: 2004. p. 6658 – 6661. DOI: 10.1021/jm049486a; (b) Wityak J, Das J, Moquin R. V, Shen Z, Lin J, Chen P, Doweyko A. M, Pitt S, Pang S, Shen D. R, Fang Q, De Fex H. F, Schieven G. L, Kanner S. B, Barrish J. C. Discovery and initial SAR of 2-amino-5-carboxamidothiazoles as inhibitors of the Src-family kinase $p56^{Lck}$, Bioorg. Med. Chem. Lett.: 2003. p. 4007 - 4010. DOI: 0.1016/j.bmcl.2003.08.054

[99] Madsen P, Kodra J. T, Behrens C, Nishimura E, Jeppesen C. B, Pridal L, Andersen B, Knudsen L. B, Aspegren C. V, Guldbrandt M, Christensen I. T, Jørgensen A. S, Ynddal L, Brand C. L, Bagger M. A, Lau J. Human Glucagon Receptor Antagonists with Thiazole Cores. A Novel Series with Superior Pharmacokinetic Properties, J. Med. Chem.: 2009. p. 2989 – 3000. DOI: 10.1021/jm8016249; (b) Kearney P, Fernandez M, Flygare J. A. Solid-Phase Synthesis of 2-Aminothiazoles, J. Org. Chem.: 1998. p. 196 - 200. DOI: 10.1021/jo971542a

[100] Cheng X, Yoshida H, Raoofi D, Saleh S, Alborzinia H, Wenke F, Go®hring A, Reuter S, Mah N, Fuchs H, Navarro M. A. A, Adjaye J, Gul S, Utikal J, Mrowka R, Wo®lfl S. Ethyl 2-((4-Chlorophenyl)amino)thiazole-4-carboxylate and Derivatives Are Potent Inducers of Oct3/4, J. Med. Chem.: 2015. p. 5742 - 5750. DOI: 10.1021/acs.jmedchem.5b00226

[101] Hencken C. P, Brando L. J, Bordon C, Stohler R, Mott B. T, Yolken R, Posner G. H, Woodard L. E. Thiazole, Oxadiazole, and Carboxamide Derivatives of Artemisinin are Highly Selective and Potent Inhibitors of Toxoplasma gondii, J. Med. Chem.: 2010. p. 3594 -3601. DOI: 10.1021/jm901857d

[102] Kumar Y, Green R, Borysko K. Z, Wise D. S, Wotring L. L, Townsend L. B. Synthesis of 2,4-disubstituted thiazoles and selenazoles as potential antitumor and antifilarial agents. 1. Methyl 4-(isothiocyanatomethyl)thiazole-2-carbamates, -selenazole-2-carbamates, and related derivatives, J. Med. Chem.: 1993. p. 3843 – 3848. DOI: 10.1021/jm00076a012

[103] Chimenti F, Secci D, Bolasco A, Chimenti P, Granese A, Carradori S, Yanez M, Orallo F, Sanna M. L, Gallinella B, Cirilli R. Synthesis, Stereochemical Separation, and Biological Evaluation of Selective Inhibitors of Human MAO-B: 1-(4-Arylthiazol-2-yl)-2-(3-methylcyclohexylidene)hydrazines, J. Med. Chem.: 2010. p. 6516 - 6520. DOI: 10.1021/jm100120s

[104] Mayhoub A. S, Khaliq M, Kuhn R. J, Cushman M. Design, Synthesis, and Biological Evaluation of Thiazoles Targeting Flavivirus Envelope Proteins, J. Med. Chem.: 2011. p. 1704 - 1714. DOI: 10.1021/jm1013538

[105] Dineen T. A, Weiss M. M, Williamson T, Acton P, Khan S. B, Bartberger M. D, Brown J, Chen K, Cheng Y, Citron M, Croghan M. D, Dunn R. T, Esmay J, Graceffa R. F, Harried S. S. Hickman D, Hitchcock S. A, Horne D. B, Huang H, Ampiah R. I, Judd T, Keller M. R, Kreiman C. R, La D. S, Li V, Lopez P, Louie S, Monenschein H, Nguyen T. T, Pennington L. D, San Miguel T, Sickmier E. A, Vargas H. M, Wahl R. C, Wen P. H, Whittington D. A, Wood S, Xue Q, Yang B. H, Patel V. F, Zhong W. Design and Synthesis of Potent, Orally Efficacious Hydroxyethylamine Derived β-Site Amyloid Precursor Protein Cleaving Enzyme (BACE1) Inhibitors. J. Med. Chem: 2012. p. 9025 - 9044. DOI: 10.1021/jm300118s

[106] Ghosh B, Antonio T, Zhen J, Kharkar P, Reith M. E. A, Dutta A. K.. Development of (S)-N6-(2-(4-(Isoquinolin-1-yl)piperazin-1-yl)ethyl)-N6-propyl-4,5,6,7-tetrahydrobenzo[d]-thiazole-2,6-diamine and Its Analogue as a D3 Receptor Preferring Agonist: Potent in Vivo Activity in Parkinson's Disease Animal Models, J. Med. Chem.: 2010. p. 1023 - 1037. DOI: 10.1021/jm901184n

[107] Das J, Chen P, Norris D, Padmanabha R, Lin J, Moquin R. V, Shen Z, Cook L. S, Doweyko A. M, Pitt S, Pang S, Shen D. R, Fang Q, de Fex H. F, McIntyre K. W, Shuster D. J, Gillooly K. M, Behnia K, Schieven G. L, Wityak J, Barrish J. C. 2-Aminothiazole as a Novel Kinase Inhibitor Template. Structure−Activity Relationship Studies toward the Discovery of N-(2-Chloro-6-methylphenyl)-2-[[6-[4-(2-hydroxyethyl)-1- piperazinyl)]-2-methyl-4-pyrimidinyl]amino)]-1,3-thiazole-5-carboxamide (Dasatinib, BMS-354825) as a Potent pan-Src Kinase Inhibitor, J. Med. Chem.: 2006. p. 6819 - 6832. DOI: 10.1021/jm060727j

[108] Hinklin R. J, Boyd S. A, Chicarelli M. J, Condroski K. R, DeWolf W. E, Lee P. A, Lee W, Singh A, Thomas L, Voegtli W. C, Williams L, Aicher T. D. Identification of a New Class of Glucokinase Activators through Structure-Based Design, J. Med. Chem.: 2013. p.7669 – 7678. DOI: 10.1021/jm401116k ; (b) Home P. D, Pacini G. Hepatic dysfunction and insulin insensitivity in type 2 diabetes mellitus: a critical target for insulin-sensitizing agents, Diabetes Obes. Metab.: 2008. p. 699 – 718. DOI: 10.1111/j.1463-1326.2007.00761.x; (c) Rizza R. A. Pathogenesis of Fasting and Postprandial Hyperglycemia in Type 2 Diabetes: Implications for Therapy, Diabetes: 2010. p. 2697 - 2707. DOI: 10.2337/db10-1032

[109] Xiao J, Marugan J. J, Zheng W, Titus S, Southall N, Cherry J. J, Evans M, Androphy E. J, Austin C. P, Discovery, Synthesis, and Biological Evaluation of Novel SMN Protein Modulators, J. Med. Chem.: 2011. p. 6215 – 6233. DOI: 10.1021/jm200497t; (b) Wirth B, Riessland M, Hahnen E. Drug discovery for spinal muscular atrophy, Expert Opin. Drug Discovery: 2007. p. 437 – 451. DOI:10.1517/17460441.2.4.437; (c) Nicolaou K. C, He Y, Roschangar F, King N. P, Vourloumis D, Li T, Total Synthesis of Epothilone E and Analogues with Modified Side Chains through the Stille Coupling Reaction, An-

gew. Chem. Int. Ed. Engl.: 1998. p. 84 - 87. DOI: 10.1002/(SICI)1521-3773(19980202)37:1/2<84::AID-ANIE84>3.0.CO;2-V

[110] Rosa M. D, Unge J, Motwani H. V, Rosenquist Å, Vrang L, Wallberg H, Larhed M. Synthesis of P1 -Functionalized Macrocyclic Transition-State Mimicking HIV-1 Protease Inhibitors Encompassing a Tertiary Alcohol, J. Med. Chem.: 2014. p. 6444 – 6457. DOI: 10.1021/jm500434q

[111] Jung F. H, Pasquet G, Brempt C. L, Lohmann J. M, Warin N, Renaud F, Germain H, De Savi C, Roberts N, Johnson T, Dousson C, Hill G. B, Mortlock A. A, Heron N, Wilkinson R. W, Wedge S. R, Heaton S. P, Odedra R, Keen N. J, Green S, Brown E, Thompson K, Brightwell S. Discovery of Novel and Potent Thiazoloquinazolines as Selective Aurora A and B Kinase Inhibitors, J. Med. Chem.: 2006. p. 955 – 970. DOI: 10.1021/jm050786h; (b) Andersen C. B, Wan Y, Chang J. W, Riggs B, Lee C, Liu Y, Sessa F, Villa F, Kwiatkowski N, Suzuki M, Nallan L, Heald R, Musacchio A, Gray N. S, Discovery of Selective Aminothiazole Aurora Kinase Inhibitors, ACS, Chem. Biolog.: 2008. p. 180 – 192. DOI: 10.1021/cb700200w; (c) Neaz M. M, Muddassar M, Pasha F, Cho S. J, Structural studies of B-type Aurora kinase inhibitors using computational methods, Acta Pharmacologica Sinica: 2010. p. 244 – 258. DOI: 10.1038/aps.2009.188; (d) Elrazaz E. Z, Serya R. A. T, Ismail N. S. M, Abou El Ella D. A, Abouzid K. A. M, Thieno[2,3-d]pyrimidine based derivatives as kinase inhibitors and anticancer agents, Future J. Pharm. Sciences. 2015. doi:10.1016/j.fjps.2015.09.001 (in press).

[112] Palmer J. T, Bryant C, Wang D. X, Davis D. E, Setti E. L, Rydzewski R. M, Venkatraman S, Tian Z. Q, Burrill L. C, Mendonca R. V, Springman E, McCarter J, Chung T, Cheung H, Janc J. W, McGrath M, Somoza J. R, Enriquez P, Yu Z. W, Strickley R. M, Liu L, Venuti M. C, Percival M. D, Falgueyret J. P, Prasit P, Oballa R, Riendeau D, Young R. N, Wesolowski G, Rodan S. B, Johnson C, Kimmel D. B, Rodan G. Design and Synthesis of Tri-Ring P3 Benzamide-Containing Aminonitriles as Potent, Selective, Orally Effective Inhibitors of Cathepsin K, J. Med. Chem.: 2005. p. 7520 - 7534. DOI: 10.1021/jm058198r

[113] Haffner C. D, Becherer J. D, Boros E. E, Cadilla R, Carpenter T, Cowan D, Deaton D. N, Guo Y, Harrington W, Henke B. R, Jeune M. R, Kaldor I, Milliken N, Petrov K. G, Preugschat F, Schulte C, Shearer B. G, Shearer T, Smalley T. L, Stewart Jr. E. L, Stuart J. D, Ulrich J. C, Discovery, Synthesis, and Biological Evaluation of Thiazoloquin(az)olin(on)es as Potent CD38 Inhibitors, J. Med. Chem.: 2015. p. 3548 -3571. DOI: 10.1021/jm502009h

[114] Kini G. D, Anderson J. D, Sanghvi Y. S, Lewis A. F, Smee D. F, Revankar G. R, Robins R. K, Cottam H. B, Synthesis and antiviral activity of certain guanosine analogs in the thiazolo[4,5-d]pyrimidine ring system, J. Med. Chem.: 1991. p. 3006 – 3010. DOI: 10.1021/jm00114a008; (b) Nagahara K, Anderson J. D, Kini G. D, Dalley N. K, Larson S. B, Smee D. F, Jin A, Sharma B. S, Jolley W. B, Robins R. K, Cottam H. B. Thiazolo[4,5-d]pyrimidine nucleosides. The synthesis of certain 3-.beta.-D-ribofuranosylthiazolo[4,5-d]pyrimidines as potential immunotherapeutic agents, J. Med. Chem.:

1990. p. 407 – 415. DOI: 10.1021/jm00163a064 (c) Webber S. E, 3, 5-Disubstituted and 3,5,7-trisubstituted-3h-oxazolo and 3h-thiazolo [4,5-d]pyrimidin-2-one compounds and prodrugs, WO2006066080: 2006.

[115] Robins R. K. Cottam H. B. Antiviral antitumor antimetastatic immune system enhancing nucleosides and nucleotides, WO198905649: 1987.

[116] Lewis A. F, Revankar G. R, Fennanwald S. M, Huffman J. H, Rando R. F. Thiazolo[4,5-d]pyrimidines. Part I. synthesis and anti-human cytomegalovirus (HCMV) activity in vitro of certain alkyl derivatives, J. Heterocyclic Chem.: 1995. p. 547 - 556. DOI: 10.1002/jhet.5570320230

[117] Carson D Thiazolopyrimidines useful as tnf$g inhibitors, WO2000/69861: 2000.

[118] Beck J. P, Curry M. A, Chorvat R. J, Fitzgerald L. W, Gilligan P. J, Zaczek R, Trainor G. L. Thiazolo[4,5-d]pyrimidine thiones and -ones as corticotropin-releasing hormone (CRH-R1) receptor antagonists, Bioorg. Med. Chem. Lett.: 1999. p. 1185 – 1188. DOI: 10.1016/S0960-894X(99)00159-6

[119] Fahmy H. T. Y, Rostom S. A. F, Saudi M. N, Zjawiony J. K, Robins D. J. Synthesis and in vitro evaluation of the anticancer activity of novel fluorinated thiazolo[4, 5-d]pyrimidines, Arch. Pharm. Pharm. Med. Chem.: 2003. p. 216 - 225. DOI: 10.1002/ardp.200300734

[120] Bekhit A. A, Fahmy H. T. Y, Rostom S. A. F, Baraka A. M, Design and synthesis of some substituted 1H-pyrazolyl-thiazolo[4,5-d]pyrimidines as anti-inflammatory–antimicrobial Agents, Eur. J. Med. Chem.: 2003. p. 27 - 36. DOI: 10.1016/S0223-5234(02)00009-0

[121] Baxter A, Cooper A, Kinchin E, Moakes K, Unitt J, Wallace A. Hit-to-Lead studies: The discovery of potent, orally bioavailable thiazolopyrimidine CXCR2 receptor antagonists, Bioorg. Med. Chem. Lett.: 2006. p. 960 – 963. DOI: 10.1016/j.bmcl.2005.10.09

[122] Binnun E. Substituted Thiazolo [4,5-d]pyrimidines as protein kinase inhibitors, US20070185139: 2007.

[123] Sircar J. C, Suto M. J, Scott M. E, Dong M. K, Gilbertsen R. B. Inhibitors of human purine nucleoside phosphorylase. Synthesis, purine nucleoside phosphorylase inhibition, and T-cell cytotoxicity of 2,5-diaminothiazolo[5,4-d]pyrimidin-7(6H)-one and 2,5-diaminothiazolo[4,5-d]pyrimidin-7(6H)-one. Two thioisosteres of 8-aminoguanine, J. Med. Chem.: 1986. p. 1804 - 1806. DOI: 10.1021/jm00159a045

[124] Ohnishi H. Thiazolopyrimidine derivative, JP82/00046 : 1982.

[125] Xing S. X, Kemnitzer W. F, Sirisoma N. S, Zhang H. N-aryl-isoxazolopyrimidin-4-amines and related compounds as activators of caspases and inducers of apoptosis and the use, WO2008/057402 : 2008.

[126] Walters I, Austin C, Austin R, Bonnert R, Cage P, Christie M, Ebden M, Gardiner S, Grahmes C, Hill S, Hunt F, Jewell R, Lewis S, Martin I, Nicholls D, Robinson D. SAR

studies on thiazolo[4,5-d]pyrimidine based CXCR2 antagonists involving a novel tandem displacement reaction, Bioorg. Med. Chem. Lett.: 2007. p. 2731 - 2734. DOI: 10.1016/j.bmcl.2007.02.080

[127] Sugihara Y. Thiazolopyrimidine derivative, EP1731523: 2006.

[128] Kung P, Meng J. J. Fused 2-amino pyrimidine compounds and their use for the treatment of cancer, WO2008/059368: 2008.

[129] Nishino T, Kobayashi T, Shinohara Y. Xanthine oxidase inhibitor, WO2007/004688: 2007.

[130] Harnden M. R, Hurst D. T. The Chemistry of Pyrimidinethiols. III. The Synthesis of Some Substituted Pyrimidinthiols and Some Thiazolo[5,4-D]pyrimidines, Aust. J. Chem.: 1990. p. 55 - 62. DOI: 10.1071/CH9900055

[131] Hurst D. T, Atcha S, Marshall K. L, The Synthesis of Some Thiazolo- and Oxazolo[5,4-d]pyrimidines and Pyrimidinylureas, Aust. J. Chem.: 1991. p. 129 - 134. DOI: 10.1071/CH9910129

[132] Miller D. J, Ravikumar K, Shen H, Suh J, Kerwin S. M, Robertus J. D, Structure-Based Design and Characterization of Novel Platforms for Ricin and Shiga Toxin Inhibition, J. Med. Chem.: 2002. p. 90 - 98. DOI: 10.1021/jm010186s

[133] Gomtsyan A, Didomenico S, Lee C, Matulenko M. A, Kim K, Kowaluk E. A, Wismer C. T, Mikusa J, Kohlhaas H. Y. K, Jarvis M. F, Bhagwat S. S. Design, Synthesis, and Structure−Activity Relationship of 6-Alkynylpyrimidines as Potent Adenosine Kinase Inhibitors, J. Med. Chem.: 2002. p. 3639 - 3648. DOI: 10.1021/jm020049a

[134] Suzuki E, Sugiura S, Naito T, Inoue S, Studies on Pyrimidine Derivatives. X, Chem. Pharm. Bull.: 1968. p. 750 - 755. DOI: org/10.1248/cpb.16.750

[135] Benedek-Vamos M, Promel R. Synthesis of pyrimidines and thiazolo[5,4-d]pyrimidines III. The N.M.R. Spectrum of thiazolo[5,4-d]pyrimidine, Tetrahedron Lett.: 1969. p. 1011 – 1014. DOI: 10.1016/S0040-4039(01)97724-X

[136] Childress S. J, Mckee R. L. Thiazolopyrimidines, J. Am. Chem. Soc.: 1951. p. 3862 - 3864. DOI: 10.1021/ja01152a090

[137] Senga K, Ichiba M, Kanazawa H, Nishigaki S, Synthesis and properties of thiazolo[5,4-d]pyrimidine 1-oxides, J. Heterocycl. Chem.: 1982. p. 77 - 80. DOI: 10.1002/jhet.5570190113

[138] Yuan D. K, Li Z. M, Zhao W. G, Synthesi of Thiazolo[5,4-d]pyrimindine-2-thiol and Its Derivatives, Chinese J. Org. Chem.: 2003. p. 1155 - 1158. DOI: sioc-journal.cn/Jwk_yjhx/EN/Y2003/V23/I10/1155

[139] Liu J, Patch R. J, Schubert C, Player M. R, Single-Step Syntheses of 2-Amino-7-chlorothiazolo[5,4-d]pyrimidines: Intermediates for Bivalent Thiazolopyrimidines, J. Org. Chem.: 2005. p. 10194 - 10197. DOI: 10.1021/jo0517702

[140] Ahmed A. F. S. Synthesis of Bicyclic Pyrimidine Nucleoside Derivatives, J. Chem. Res. (S): 1998. p. 697 - 697. DOI: 10.1039/A805968C

[141] Freeman F, Kim D. S. H. L. Reaction of aminopropanedinitrile 4-methylbenzenesulfonate (aminomalononitrile p-toluenesulfonate (tosylate)) with isothiocyanates, J. Org. Chem.: 1991. p. 4645 - 4648. DOI: 10.1021/jo00015a015

[142] Hennen W. J, Hinshaw B. C, Riley T. A, Wood S. G, Robins R. K. Synthesis of 4-substituted 5-amino-2-(.beta.-D-ribofuranosyl)thiazoles and 4-substituted 5-amino-2-(.beta.-D-ribofuranosyl)selenazoles, and their respective conversion into 2-(.beta.-D-ribofuranosyl)thiazolo[5,4-d]pyrimidines and 2-(.beta.-D-ribofuranosyl)selenazolo[5,4-d]pyrimidines. A new synthesis of tiazofurin and selenazofurin, J. Org. Chem.: 1985. p. 1741 - 1746. DOI: 10.1021/jo00210a033

[143] Beck J. P. Thiazolo[4,5-d]pyrimidines and pyridines as corticotropin releasing factor (crf) antagonists, WO1999/51608: 1999.

[144] Khaury A. M, Bayouki E, Basyouni W. M. New Thiazolo[5,4-d]pyrimidines with Molluscicidal Properties, Bull. Chem. Soc. Jpn.: 1988. p. 3794 - 3796. DOI: org/10.1246/bcsj.61.3794

[145] Corrao S. L, Macielag M. J, Turchi I. J. Rearrangement of 4-(aminothiocarbonyl)oxazoles to 5-aminothiazoles. Synthetic and MINDO/3 MO studies, J. Org. Chem.: 1990. p. 4484 - 4487. DOI: 10.1021/jo00301a056

[146] Senga K, Sato J, Nishigaki S. A New, Facile Synthesis of Oxazolo [5, 4-d] pyrimidines and Their Conversion into Thiazolo [5, 4-d] pyrimidines, Chem. Pharm. Bull.: 1978. p. 765 - 769. DOI: doi.org/10.1248/cpb.26.765

[147] Baker J. A, Chatfield P. V. Synthesis of derivatives of thiazolo[4,5-d]pyrimidine, J. Chem. Soc. (C): 1969. 603 - 606. DOI: 10.1039/J39690000603

[148] Mi-Yeon Jang, Steven De Jonghe, Ling-Jie Gao, Bart Vanderhoydonck, Piet Herdewijn, Novel thiazolopyrimidines, British patent application GB09043530.

[149] Bodeker J, Courault K, Kockritz A, Kockritz P. Reaktionen von Heteroaryliminotriphenylphosphoranen mit Heterokumulenen: Synthese und Cycloadditionen α-N-Heteroaryl-substituierter Carbodiimide, J. Prakt. Chem.: 1983. p. 463 – 474. DOI: 10.1002/prac.19833250316

[150] Gewald K. Heterocyclen aus CH-aciden Nitrilen. VI. Reaktion von methylenaktiven Nitrilen mit Senfölen und Schwefel, J. Prakt. Chem.: 1966. p. 26 - 30. DOI: 10.1002/prac.19660320104

[151] Stuckwisch C. G. Azomethine Ylids, Azomethine Imines, and Iminophosphoranes in Organic Syntheses, Synthesis: 1973. p. 469 – 483. DOI: 10.1055/s-1973-22235

[152] Molina P, Arques A, Vinader M. V. Fused pyrimidines by a tandem aza-Wittig/electrocyclic ring closure strategy: synthesis of pyrazolo[3,4-d]pyrimidine, [1,2,3]triazo-

lo[4,5-d]pyrimidine, and thiazolo[4,5-d]pyrimidine derivatives, J. Org. Chem.: 1988. p. 4654 - 4663. DOI: 10.1021/jo00255a003

[153] Kale M. G, Raichurkar A, Waterson S. H. P. D, McKinney D, Manjunatha M. R, Kranthi U, Koushik K, Jena L. K, Shinde V, Rudrapatna S, Barde S, Humnabadkar V, Madhavapeddi P, Basavarajappa H, Ghosh A, Ramya V. K, Guptha S, Sharma S, Vachaspati P, Kumar K. N. M, Giridhar J, Reddy J, Panduga V, Ganguly S, Ahuja V, Gaonkar S, Kumar C. N. N, Ogg D, Tucker J. A, Sjodin P. A. B, de Sousa S. M, Sambandamurthy V. K, Ghorpade S. R. Thiazolopyridine Ureas as Novel Antitubercular Agents Acting through Inhibition of DNA Gyrase B, J. Med. Chem.: 2013. p. 8834 - 8848. DOI: 10.1021/jm401268f

[154] Bebernitz G. R, Beaulieu V, Dale B. A, Deacon R, Duttaroy A, Gao J, Grondine M. S, Gupta R. C, Kakmak M, Kavana M, Kirman L. C, Liang J, Maniara W. M, Munshi S, Nadkarni S. S, Schuster H. F, Stams T, Denny I. St, Taslimi P. M, Vash B, Caplan S. L. Investigation of Functionally Liver Selective Glucokinase Activators for the Treatment of Type 2 Diabetes, J. Med. Chem.: 2009. p. 6142 - 6152. DOI: 10.1021/jm900839k

[155] Dolle R. E, Le Bourdonnec B, Goodman A. J, Morales G. A, Thomas C. J, Zhang W. J. Comprehensive Survey of Chemical Libraries for Drug Discovery and Chemical Biology: 2008, Comb. Chem.: 2009. p. 739 - 790. DOI: 10.1021/cc9000828

[156] Lee J, Wu X, Pasca di Magliano M, Peters E. C, Wang Y, Hong J, Hebrok M, Ding S, Cho C. Y, Schultz P. G. A Small-Molecule Antagonist of the Hedgehog Signaling Pathway, ChemBioChem: 2007. p. 1916 - 1919. DOI: 10.1002/cbic.200700403

[157] Milne J. C, Lambert P. D, Schenk S, Carney D. P, Smith J. J, Gange D. J, Jin L, Boss O, Perni R. B, Vu C. B, Bemis J. E, Xie R, Disch J. S, Ng P. Y, Nunes J. J, Lynch A. V, Yang H, Galonek H, Israelian K, Choy W, Iffland A, Lavu S, Medvedik O, Sinclair D. A, Olefsky J. M, Jirousek M. R, Elliott P. J, Westphal C. H. Small molecule activators of SIRT1 as therapeutics for the treatment of type 2 diabetes, Nature: 2007. p. 712 -716. DOI: 10.1038/nature06261 (b) Das J, Chen P, Norris D, Padmanabha R, Lin J, Moquin R. V, Shen Z, Cook L. S, Doweyko A. M, Pitt S, Pang S, Shen D. R, Fang Q, de Fex H. F, McIntyre K. W, Shuster D. J, Gillooly K. M, Behnia K, Schieven G. L, Wityak J, Barrish J. C. 2-Aminothiazole as a Novel Kinase Inhibitor Template. Structure−Activity Relationship Studies toward the Discovery of N-(2-Chloro-6-methylphenyl)-2-[[6-[4-(2-hydroxyethyl)-1- piperazinyl)]-2-methyl-4-pyrimidinyl]amino)]-1,3-thiazole-5-carboxamide (Dasatinib, BMS-354825) as a Potent pan-Src Kinase Inhibitor, J. Med. Chem.: 2006. p. 6819 - 6832. DOI: 10.1021/jm060727j

[158] Cee V.J, Frohn M, Lanman B. A, et al., Discovery of AMG 369, a Thiazolo[5,4-b]pyridine Agonist of S1P1 and $S1P_5$, ACS Med. Chem. Lett.: 2011. p. 107 - 112. DOI: 10.1021/ml100306h

[159] Smethurst C. A, Borthwick J. A, Gaines S, Watson S, Green A, Schulz M. J, Burton G, Buson A. A, Arban R. The characterization of a novel V1b antagonist lead series, Bioorg. Med. Chem. Lett.: 2011. p. 92 - 96. DOI: 10.1016/j.bmcl.2010.11.061

[160] Thomae D, Perspicace E, Hesse S, Kirsch G, Seck P. Synthesis of substituted [1,3]thiazolo[4,5-b]pyridines and [1,3]thiazolo[4,5-d, 1,2,3]triazines, Tetrahedron: 2008. p. 9309 - 9314. DOI: 10.1016/j.tet.2008.07.017

[161] Thomae D, Kirsch G, Seck P. Synthesis of Thiophene Analogues of the Tacrine Series, Synthesis: 2007. p. 1027 - 1032. DOI: 10.1055/s-2007-965944

[162] Komoriya S, Kobayashi S, Osanai K, Yoshino T, Nagata T, Haginoya N, Nakamoto Y, Mochizuki A, Nagahara T, Suzuki M, Shimada T, Watanabe K, Isobe Y, Furugoori T. Design, synthesis, and biological activity of novel factor Xa inhibitors: Improving metabolic stability by S1 and S4 ligand modification, Bioorg. Med. Chem.: 2006. p. 1309 - 1330. DOI: 10.1016/j.bmc.2005.09.056

[163] Kulkarni S. S, Newman A. H, Discovery of heterobicyclic templates for novel metabotropic glutamate receptor subtype 5 antagonists, Bioorg. Med. Chem. Lett.: 2007. p. 2987 - 2991. DOI: 10.1016/j.bmcl.2007.03.066

[164] Walczynski K, Zuiderveld O. P, Timmerman H. Non-imidazole histamine H3 ligands. Part III. New 4-n-propylpiperazines as non-imidazole histamine H3-antagonists, Eur. J. Med. Chem.: 2005. p. 15 - 23. DOI: 10.1016/j.ejmech.2004.09.010

[165] Lin R, Johnson S. G, Connolly P. J, Wetter S. K, Binnun E, Hughes T. V, Murray W. V, Pandey N. B, Moreno-Mazza S. J, Adams M, Fuentes-Pesquera A. R, Middleton S. A. Synthesis and evaluation of 2,7-diamino-thiazolo[4,5-d] pyrimidine analogues as anti-tumor epidermal growth factor receptor (EGFR) tyrosine kinase inhibitors, Bioorg. Med. Chem. Lett.: 2009. p. 2333 - 2337. DOI: 10.1016/j.bmcl.2009.02.067

[166] Singh B, Bacon E. R, Lesher G. Y, Robinson S, Pennock P. O, Bode D. C, Pagani E. D, Bentley R. G, Connell M. J, Hamel L. T, Silver P. J. Novel and Potent Adenosine 3',5'-Cyclic Phosphate Phosphodiesterase III Inhibitors: Thiazolo[4,5-b, 1,6]naphthyridin-2-ones, J. Med. Chem.: 1995. p. 2546 - 2550. DOI: 10.1021/jm00014a007

[167] Lee T, Lee D, Lee Y, Gong Y. D. Solid-Phase Synthesis of Thiazolo[4,5-b]pyridine Derivatives Using Friedländer Reaction, J. Comb. Chem.: 2010. p. 95 - 99. DOI: 10.1021/cc900147y

[168] Takahashi Y, Hashizume M, Shin K, Terauchi T, Takeda K, Hibi S, M.Tai K, Fujisawa M, Shikata K, Taguchi R, Ino M, Shibata H, Yonaga M. Design, Synthesis, and Structure–Activity Relationships of Novel Pyrazolo[5,1-b]thiazole Derivatives as Potent and Orally Active Corticotropin-Releasing Factor 1 Receptor Antagonists, J. Med. Chem.: 2012. p. 8450 - 8463. DOI: 10.1021/jm300864p

Chapter 2

Recent Advances in the Biological Importance of Rhodanine Derivatives

Amit B. Patel and Premlata Kumari

Additional information is available at the end of the chapter

http://dx.doi.org/10.5772/62835

Abstract

Heterocyclic compounds are an important part of the synthetic medicinal chemistry. They offer a high degree of structural variety and have proven to be widely useful as therapeutic agents. Heterocyclic compounds play an important role in the biological processes. They are widespread as natural products. Heterocyclic compounds are widely found in nature categorically in plant alkaloids, nucleic acids, anthocyanins, and flavones. They are also present as in chlorophyll and hemoglobin. Additionally, some proteins, hormones, and vitamins also contain aromatic heterocyclic system. Heterocycles have huge potential as the most promising molecules as lead structures for the design of new drugs. About one half of over 6 million compounds recorded so far in chemical abstracts are heterocyclic. The proposed book chapter entitled, *Recent Advances in the Biological Importance of Rhodanine Derivatives* gives an outline of importance and applications of the various rhodanine derivatives in medicinal chemistry from 2004 to 2014.

Keywords: Rhodanine, biological activities, structure activity relationship and selectivity of rhodanine derivatives

1. Introduction

Rhodanine is a five-membered heterocyclic molecule containing a thiazole nucleus with thioxo group on second carbon and carbonyl group on fourth carbon. It was first discovered in 1877 by Marceli Nencki, who named it "Rhodaninsaure." Structural modifications of rhodanine derivatives (Figure 1) constantly result in compounds with a broad spectrum of pharmacological activities [1, 2]. Rhodanine derivatives recently have grabbed the attention of researchers because of their broad range of pharmacological activities. Since past 10 years, the number of scientific publications and patents describing a plenty of the different biological activities of rhodanine-based compounds is increasing continuously (Figure 2). It has been reached at the peak in 2014 with 461 publications. A majority of the biologically active rhodanines are 5-arylmethylidenerhodanines (Figure 1), which contain the exocyclic double bond. Because the latter is conjugated to the carbonyl group at position 4 of the rhodanine ring, such compounds are electrophilic and potentially reactive due to possible Michael addition of the nucleophilic protein residues to the exocyclic double bond [3–5].

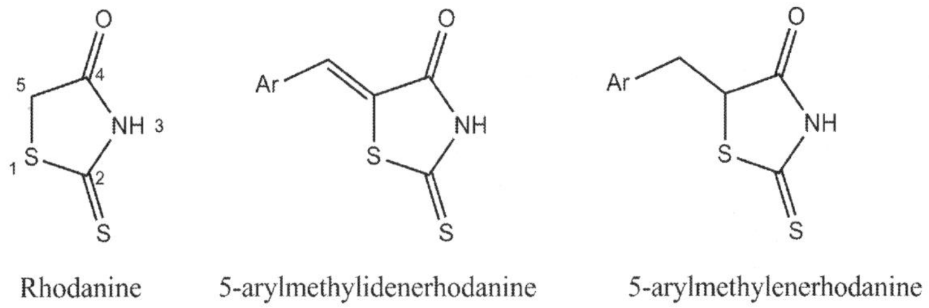

Figure 1. Chemical structures of the important rhodanine-based derivatives.

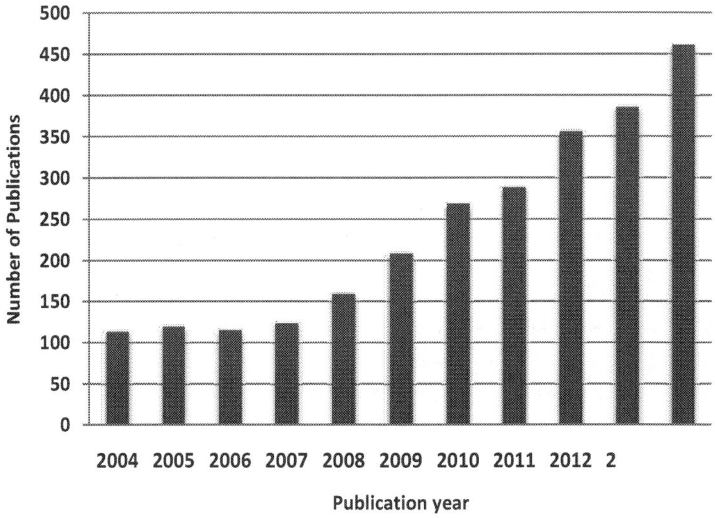

Figure 2. SciFinder search for recent publications, including biological activity of rhodanines sorted by year, as determined on 10 August 2015.

Rhodanine have been found to possess various biological activities, such as antidiabetic, antibacterial, antifungal, anti-infective, pesticidal, antimycobacterial, antineoplastic, and so on [6–19]. They also exhibit antitubercular, anti–human immunodeficiency virus (HIV), and antimalarial activities. Due to the various possibilities of structural derivatization of the

http://dx.doi.org/10.5772/62835

rhodanine ring, their derivatives will probably remain a privileged scaffold in drug discovery [20]. We therefore want to review the biological activities, mechanism of action, structure–activity relationship (SAR), and selectivity of rhodanine derivatives against various targets in this chapter.

1.1. Antibacterial activity

Villain-Guillot et al. [21] have reported design, synthesis, and SAR of furanyl-substituted rhodanine derivatives as RNA polymerase (RNAP) inhibitors. These derivatives were found to inhibit transcription and affect growth of bacteria living in suspension or in a biofilm. The derivative (I) is found as most active among all the reported rhodanine derivatives. It inhibits the *Escherichia coli* RNAP transcription at minimum inhibition concentration of ≤10 μM. It also have high efficacy against various gram-positive bacteria, including *Staphylococcus epidermidis*.

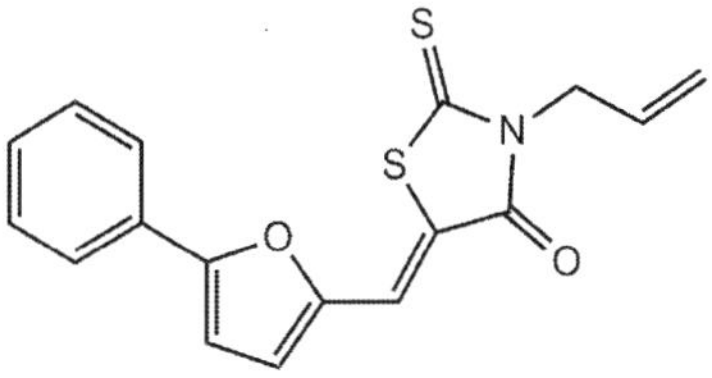

(I)

Hardej et al. [22] have synthesized a series of rhodanine derivatives containing various substituents at the N_3- and C_5-positions and tested for in vitro antibacterial activity against a panel of clinically relevant methicillin-resistant *Staphylococcus aureus* (MRSA) strains. The anti-MRSA activity of compounds II (minimum inhibitory concentration (MIC)=3.9 μg/mL) and III (MIC=1.95 μg/mL) were significantly greater than that of the reference antibiotics penicillin G (MIC=31.25 μg/mL) and ciprofloxacin (MIC=7.8 μg/mL).

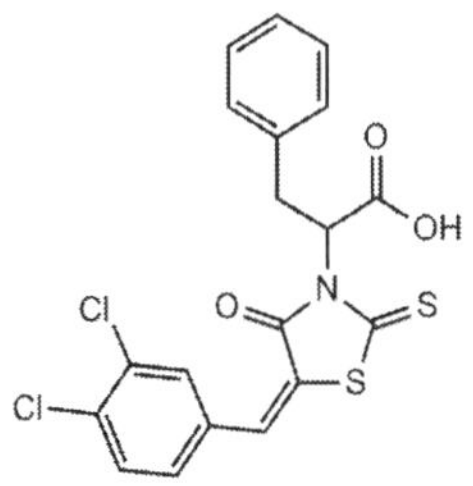

(II)

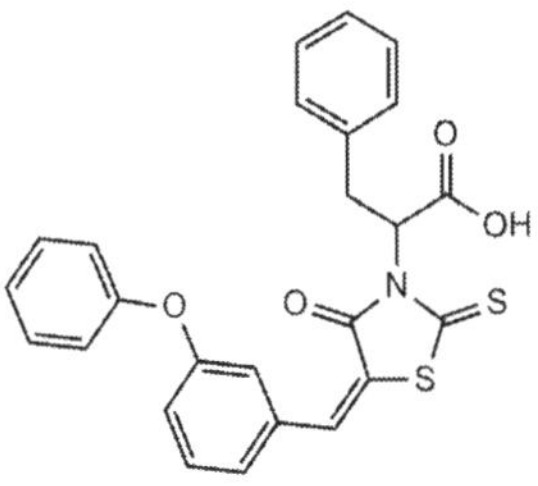

(III)

Li et al. [23] have synthesized a series of arylhydrazone derivatives bearing a rhodanine moiety and evaluated as antibacterial activity against several different strains of gram-positive bacteria, including multidrug-resistant clinical isolates. Of all the compounds tested, IV and V were identified as the most effective, with minimum inhibitory concentration values of 2–4 μg/mL against methicillin-resistant and quinolone-resistant *S. aureus*.

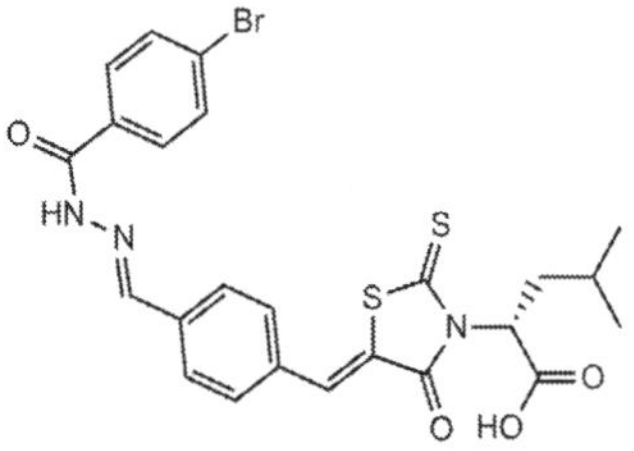

(IV)

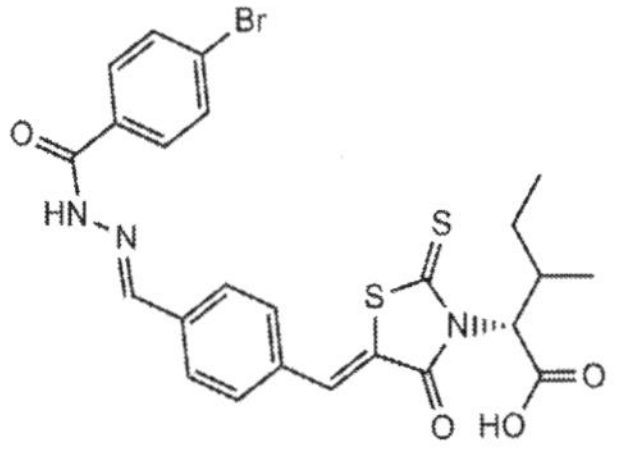

(V)

Zheng et al. [24] have synthesized three novel series of 5-aryloxypyrazole derivatives and tested for their antibacterial activity. The majority of the synthesized compounds showed potent inhibitory activity against gram-positive bacteria *S. aureus* 4220, especially against the strains of multidrug-resistant clinical isolates (MRSA3167/3506 and QRSA3505/3519). Among which, compounds VI, VII, and VIII showed the most potent levels of activity (MIC=1 μg/mL), and cytotoxic activity assay showed that the compounds tested did not affect cell viability on the human cervical (HeLa) cells at their MICs.

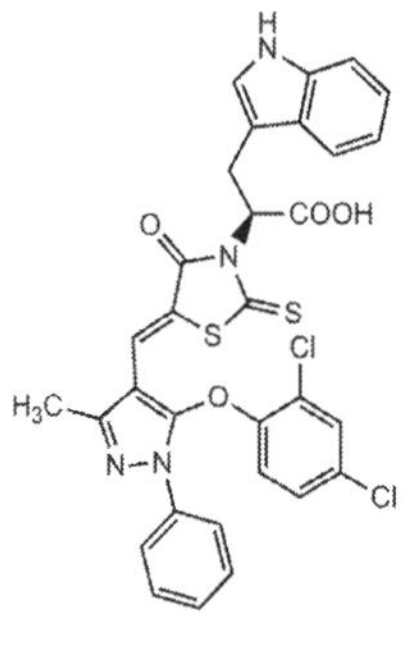

(VI)

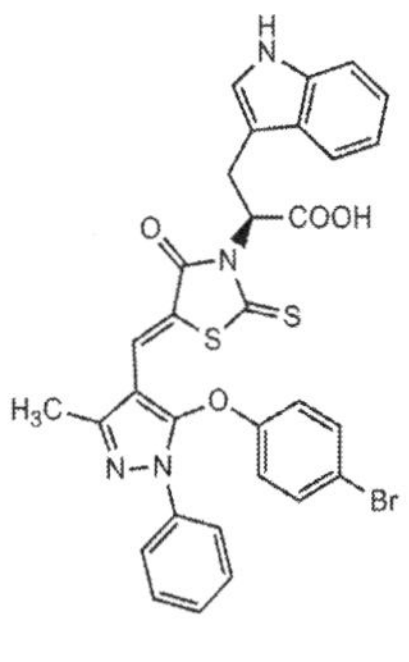

(VII)

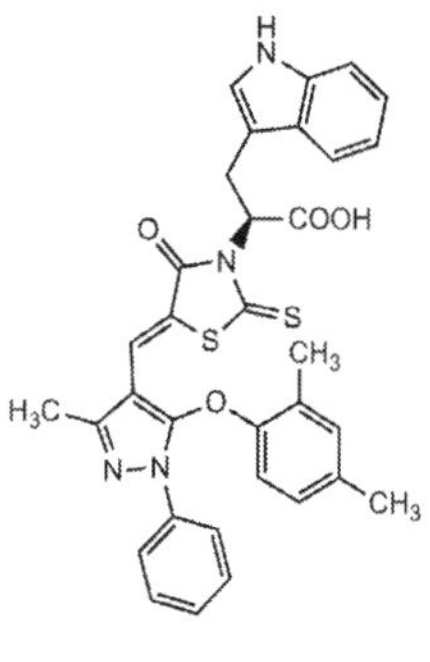

(VIII)

Xu et al. [25] synthesized pyrazole-substituted derivatives bearing rhodanine-3-fatty acid moieties and analyzed their antimicrobial activities against various gram-positive as well as gram-negative bacteria. Compound (IX) bearing a rhodanine-3-pentanoic acid displayed the most potent activity with a MIC of 2 μg/mL against MRSA.

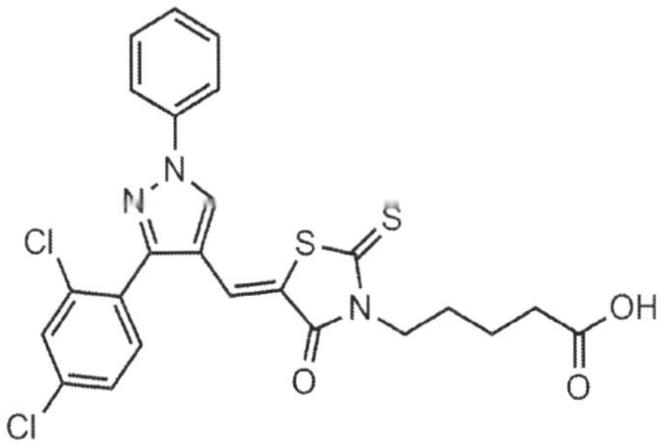

(IX)

Miao et al. [26] have synthesized a series of rhodanine-3-acetic acid derivatives and investigated for their antibacterial activity against gram-positive bacteria, including multidrug-resistant clinical isolates. The compounds X, XI, XII, XIII, XIV, and XV presented better activities against multidrug-resistant *S. aureus* than the standard drug, especially XIII with a MIC of 1 μg/mL. However, none of the compounds were active against gram-negative bacteria at 64 μg/mL.

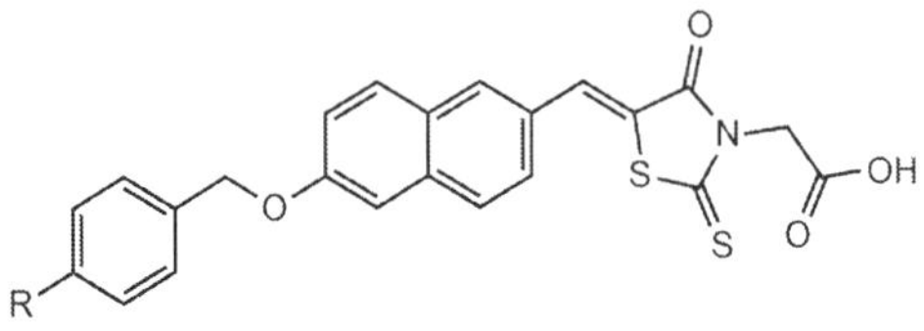

(X)-(XV)

(X)= CH_3, (XI)= OCH_3, (XII)= F,

(XIII)= Br, (XIV)= Cl, (XV)= CF_3

1.2. Antifungal activity

Orchard et al. [27] have synthesized rhodanine-3-acetic acid derivatives XVI, XVII, and XVIII inhibit *Candida albicans* PMT1 with inhibition concentration 50% (IC_{50}) values 0.17, 0.2, and 0.35 μM, respectively. These compounds could serve as useful tools for studying the effects of protein O-mannosylation and its relevance in the search for novel antifungal agents.

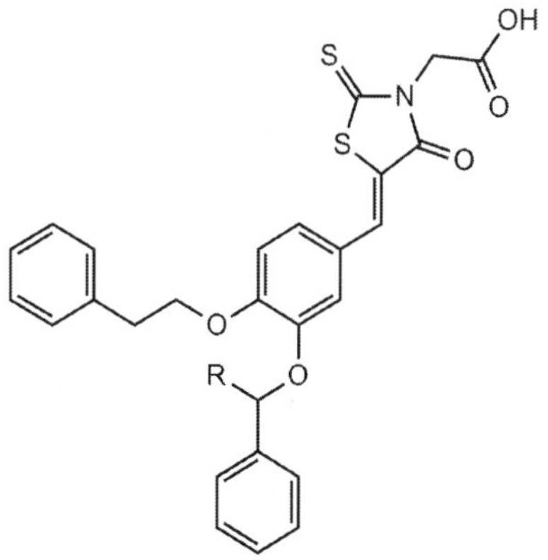

(XVI)-(XVIII)

(XVI)= CH_2OH, (XVII)= CH_3, (XVIII)= $CONH_2$

Sortino et al. [28] reported a series of benzylidene–rhodanines acting as antifungal agents. Among them, compounds XIX and XX showed to be fungicides and were the most active against *Candida genus* and *Candida neoformans*, including clinical isolates.

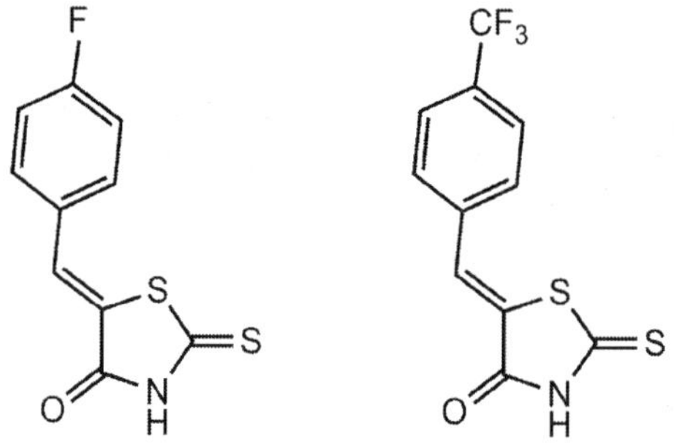

(XIX) (XX)

In an effort to develop highly potent antifungal agents, Chauhan et al. [29] have reported potent antifungal rhodanine analogs. Some derivatives XXI, XXII, and XXIII were found to be very effective (MIC=0.78 µg/mL) against *C. albicans* MTCC183. The potent compounds were further tested for in vitro anticandidal activity and amphotericin B–resistant strain of *C. albicans*.

Moreover, these analogs did not exhibit any toxicity up to MIC 3.12 μg/mL against mammalian cell line L929.

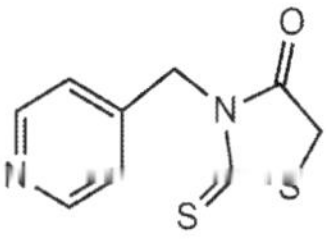

(XXI)

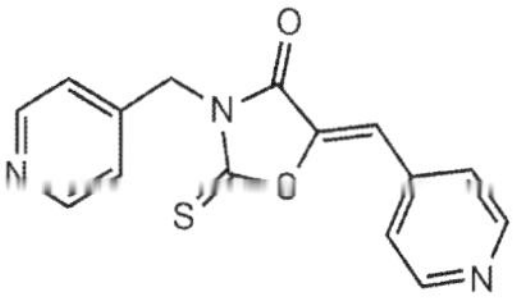

(XXII)

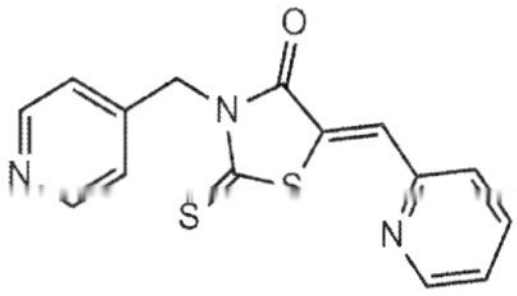

(XXIII)

Insuasty et al. [30] have synthesized several simple rhodanine derivatives and tested for their antifungal activity against 10 different fungal strains. Compound XXIV showed high activity against *Saccharomyces cerevisiae* (MIC 3.9 μg/mL) of all the tested derivatives.

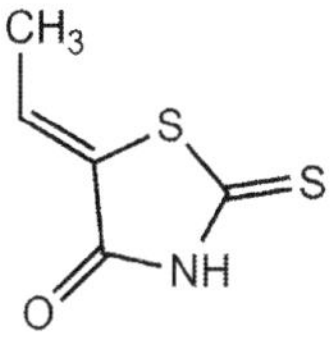

(XXIV)

1.3. Antidiabetic activity

Murugan et al. [31] illustrated simple and efficient synthesis of regio- and stereo-controlled dispiropyrrolidine derivatives of rhodanine XXV, which are found to exhibit attractive antidiabetic properties to male Wistar rats. Among the eight rhodanine compounds, particularly two compounds showed the excellent antidiabetic activity

1.4. Anticancer activity

Moorthy et al. [32] have synthesized 5-isopropylidiene derivatives of 3-dimethyl-2-thiohydantoin XXVI, 3-ethyl-2-thio-2,4-oxazolidinedione XXVII, and 5-benzilidene-3-ethyl rhodanine XXVIII, which are cytotoxic against leukemic cell line in concentration-dependent manner. The results of the trypan blue and MTT assays indicated that the compound XXVIII found to be fivefold to sevenfold more potent than XXVI and XXVII with IC_{50}<10 μM. XXVIII found to affect DNA replication by inducing a block at S phase on the basis of cell cycle analysis and tritiated thymidine assays. Moreover, the treatment of XXVIII led to increased level of reactive oxygen species (ROS) production and DNA strand breaks. This suggests the activation of apoptosis for induction of cell death.

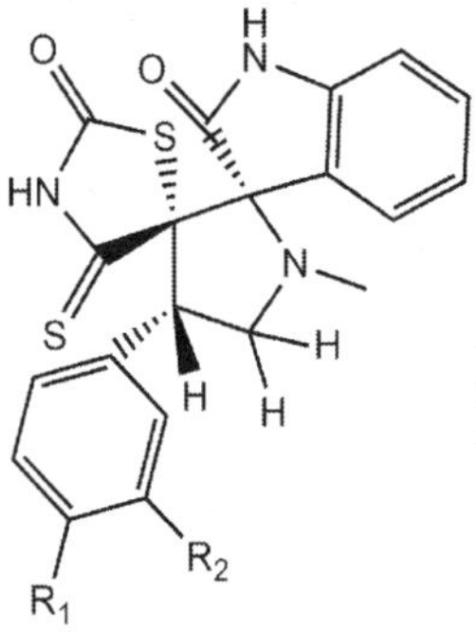

R_1=R_2= F
R_1=R_2= OCH_3

(XXV)

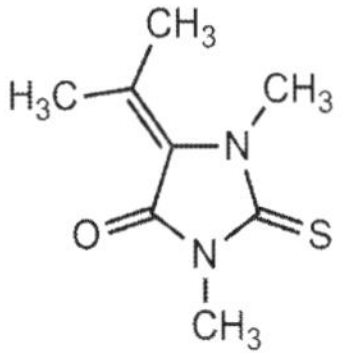

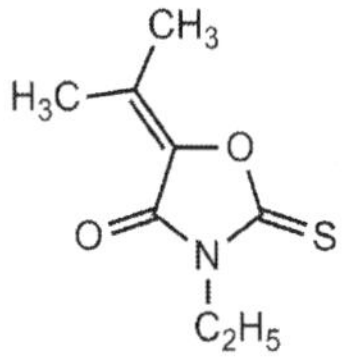

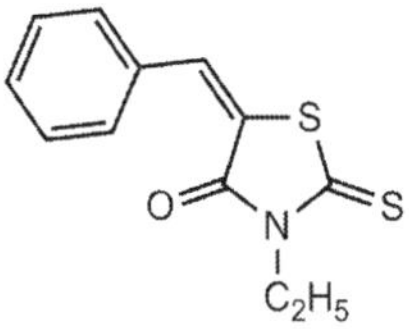

Li et al. [33] have synthesized a series of rhodanine-containing sorafenib derivatives. The in vitro pharmacological activity indicated that some of the target compounds possessed high antitumor activity against cancer cell lines, such as A549, H460, and HT29, compared to the standard drug sorafenib. The compound XXIX has displayed highest IC_{50} value of 0.8, 1.3, and 2.8 μM against A549, H460, and HT29 cell lines, respectively. The SAR data indicated that the activity strongly depends on the substitution pattern of the rhodanine motif at C-5 position.

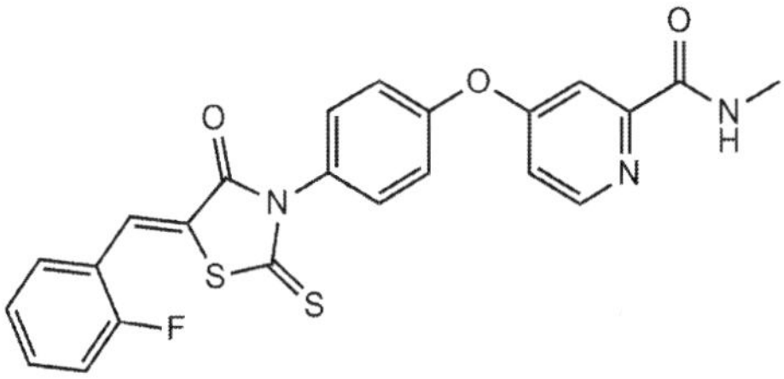

(XXIX)

Liu et al. [34] synthesized a series of dihydropyrimidinone and rhodanine derivatives and tested their tyrosinase inhibitory activity. The results showed that some of the synthesized derivatives displayed significant inhibitory activities. The SAR data indicated that the compound XXX with the presence of hydroxyethoxyl group at position 4 of phenyl ring has displayed highest tyrosinase inhibitory activity with IC_{50} value of 0.56 μM. The inhibitory effect of compound XXX on the tyrosinase was found to be irreversible. These results suggested that such compounds might be served as lead for further designing of new potential tyrosinase inhibitors.

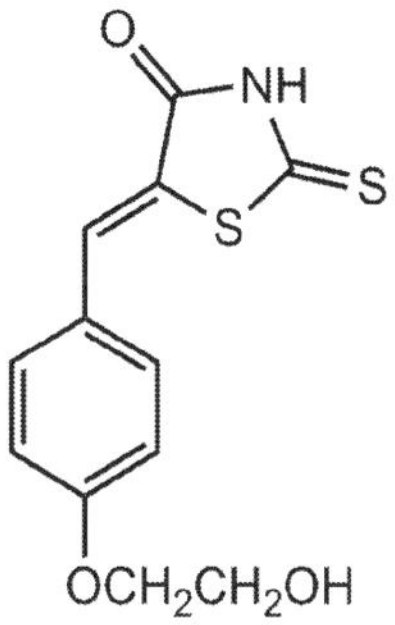

(XXX)

Min et al. [35] synthesized rhodanine derivatives, XXXI and XXXII, which inhibited protein tyrosine phosphatase type IVA, member 3 (PRL-3) enzymatic activity with IC_{50} values of 0.8 and 1.1 μM, respectively. These two derivatives highly inhibited the migration and invasion of PRL-3 overexpressing colon cancer cells. The phosphorylation recovery of known PRL-3 substrates, such as ezrin and cytokeratin, confirmed the specificity of the inhibitors on PRL-3 phosphatase activity. These compounds also selectively inhibited the PRL-3 when compared to the other phosphatases. Moreover, the derivative XXXI also found to regulate the epithelial-to-mesenchymal transition (EMT) marker proteins.

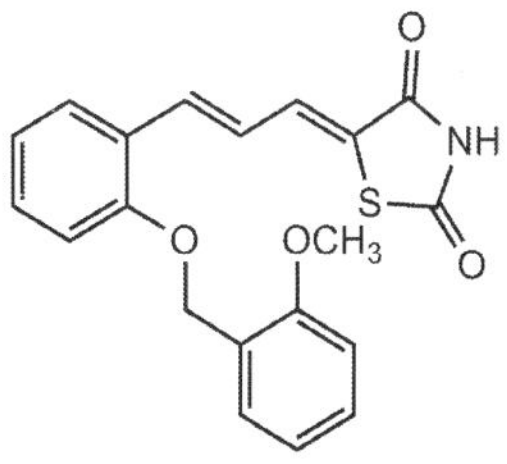

(XXXI)

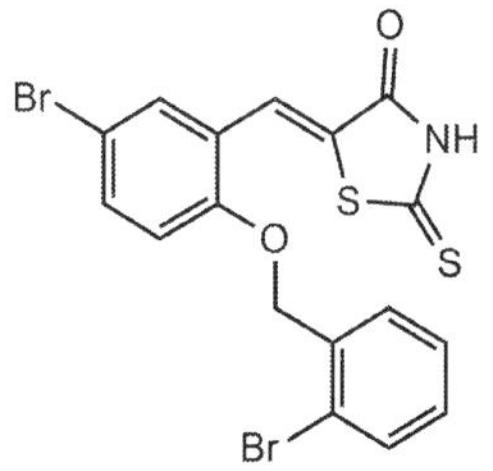

(XXXII)

1.5. Anti-HIV

Rajamaki et al. [36] have reported a novel series of rhodanine derivatives inhibiting HIV-1 integrase using virtual screening techniques. The compound XXXIII has displayed highest therapeutic index (7.0) of all the synthesized derivatives.

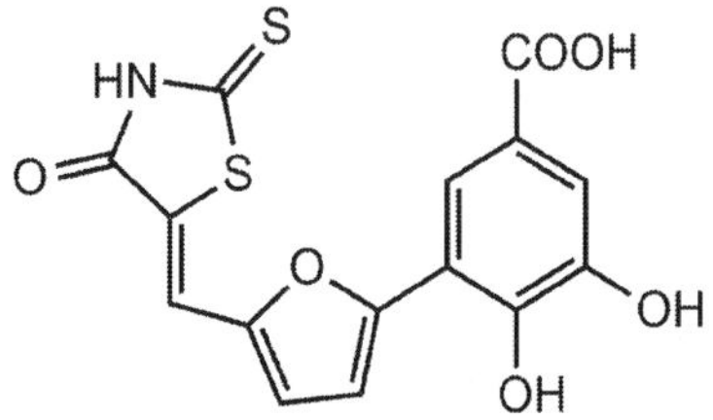

Maga et al. [37] synthesized a series of second-generation rhodanine derivatives with high inhibitory activity toward cellular DEAD (Asp-Glu-Ala-Asp) (DDX3) and HIV-1 replication using optimization protocol to the first non-nucleoside inhibitor of the adenylpyrophosphatase (ATPase) activity of human DEAD-box RNA helicase DDX3. Rationalized biological data in terms of SAR and docking simulations indicated that compound XXXIII displayed highest selectivity index (10.0) of all the synthesized rhodanine derivatives.

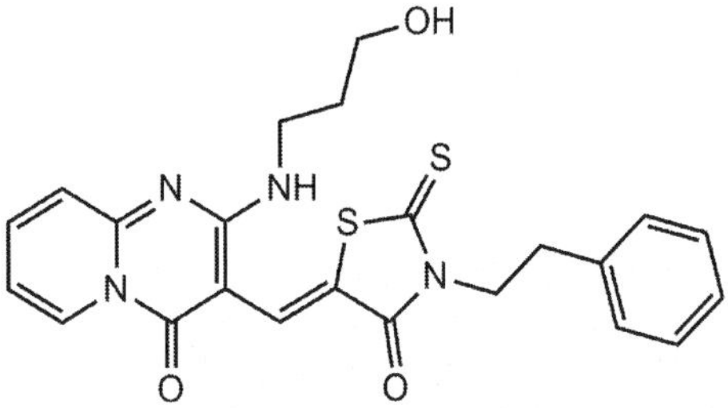

(XXXIII)

Jiang et al. [38] reported syntheses of furan-substituted rhodanine derivatives by Suzuki-Miyaura cross-coupling, followed by Knoevenagel condensation reaction. The derivatives XXXIV and XXXV have shown excellent potency against primary HIV-1 strains with effective concentration 50% (EC_{50}) at low nanomolar level of all the synthesized derivatives. The SAR data indicated that these derivatives also inhibit the HIV-1–mediated cell–cell fusion and the glycoprotein 41 (gp41) six-helix bundle formation.

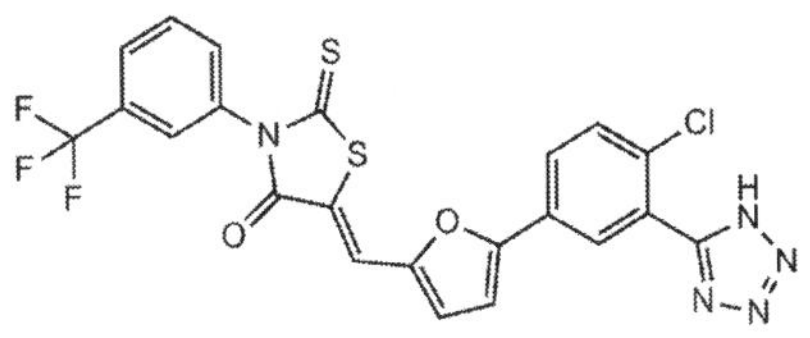

(XXXIV)

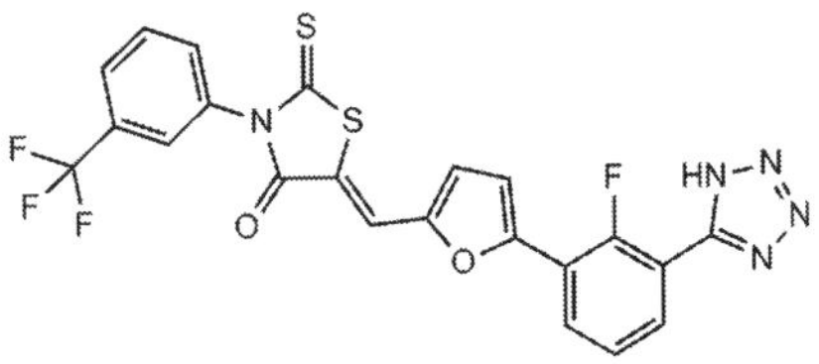

(XXXV)

1.6. Anti–hepatitis C virus activity

Talele et al. [39] reported novel allosteric inhibitors of hepatitis C virus (HCV) nonstructural protein 5B (NS5B) through a combination of structure-based virtual screening, synthesis, and SAR optimization approach. All the derivatives that exhibited IC_{50} values ranging from 7.7 to 68.0 μM were developed. Compound XXXVI, a novel rhodanine analog with NS5B inhibitory potency in the low micromolar level range may be a promising lead for future development of more potent NS5B inhibitors.

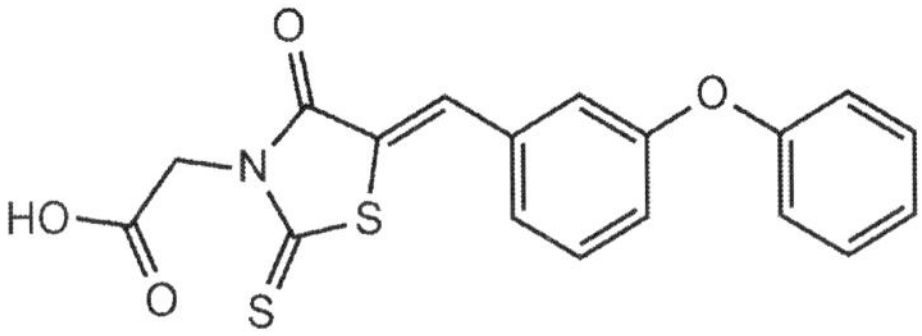

(XXXVI)

Patel et al. [40] have reported the synthesis and in vitro evaluation of anti-NS5B polymerase activity of some novel rhodanine derivatives. Depending on the nature of substituents, the tested compounds exhibited IC_{50} values ranging between 2 and 50 μM against NS5B polymerase. Analogue (XXXVII) have displayed highest IC_{50} (2.6 μM) of all the tested rhodanine derivatives.

1.7. Anti-Inflammatory agent

Cutshall et al. [41] have synthesized a series of rhodanine-based inhibitors and tested against the dual-specificity phosphatases (DSP) family member c-Jun N-terminal kinases (JNK)-stimulating phosphatase-1 (JSP-1). The SAR studies demonstrated that presence of stronger electron-withdrawing functional groups at aryl-benzylidene position provided analogs with the greatest potencies as illustrated by compound (XXXVIII). These derivatives may be useful for the treatment of inflammatory and proliferative disorders.

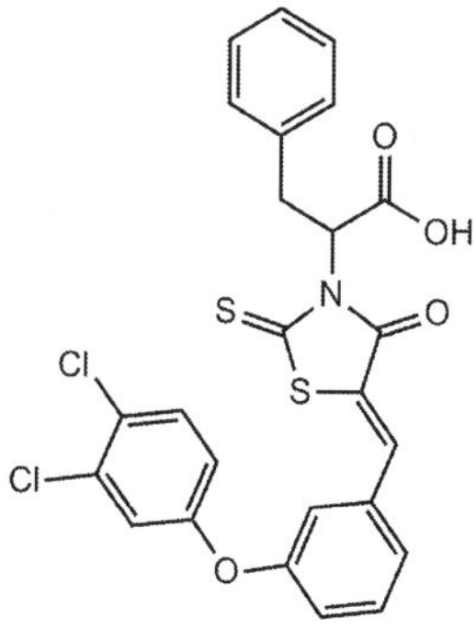

(XXXVII)

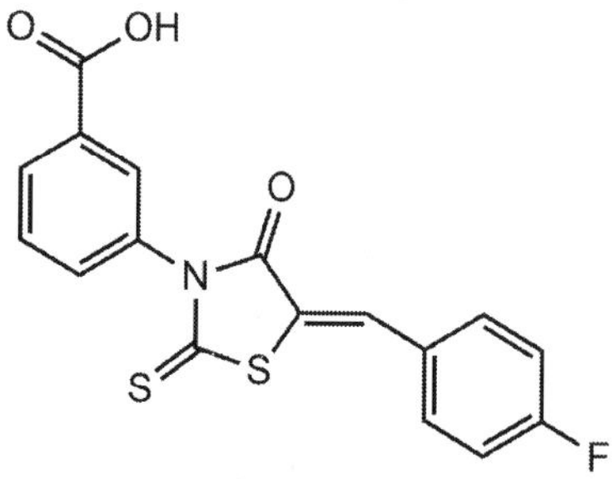

(XXXVIII)

Irvine et al. [42] have reported the in vitro anti-inflammatory activity of a novel series of rhodanine-based phosphodiesterase-4 (PDE4) inhibitors. From the SAR study, it was observed that analog XXXIX (IC_{50}=0.89 μM) and XXXX (IC_{50} 0.74 μM) displayed highest anti-inflammatory activity.

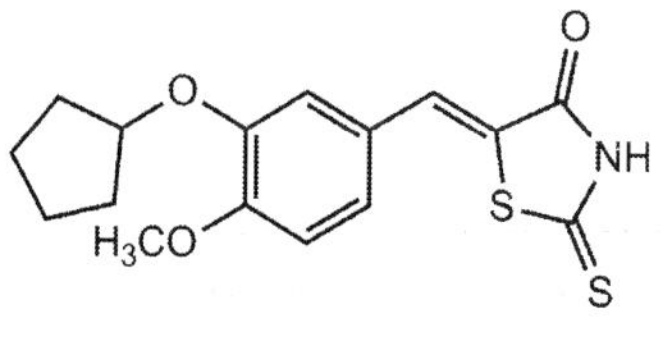

(XXXIX)

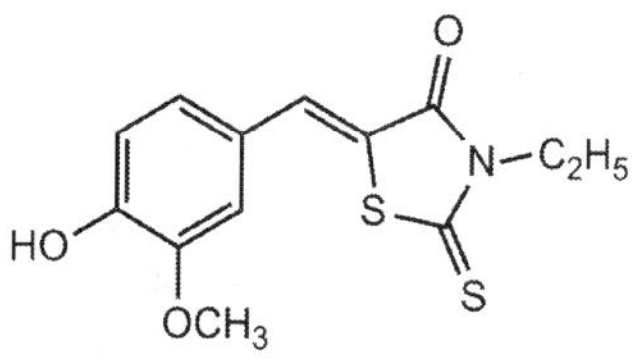

(XXXX)

http://dx.doi.org/10.5772/62835

2. Conclusion

This chapter describes rhodanine-based compounds that have been highly associated with biological activity, especially with antibacterial, antiviral, and anticancer activities. Rhodanine derivatives have attracted huge attention of millions of chemists and biologist in recent time because of their wide range of pharmacological activities and therefore, further improved protocol with better observation is still under progress. To conclude, rhodanines will probably remain a privileged scaffold in drug discovery due to their wide spectrum of pharmacological activity and the different possibilities of structural modification, which enable potent and selective drugs to be developed.

Acknowledgements

The authors are thankful to Principal, Government College Daman and Administration of Daman and Diu as well as S. V. National Institute of Technology, Surat, for providing all the facilities and wonderful working environment.

Author details

Amit B. Patel[1] and Premlata Kumari[2*]

*Address all correspondence to: premlatakumari1@gmail.com

1 Department of Chemistry, Government College Daman, Daman (U.T.), India

2 Department of Applied Chemistry, S. V. National Institute of Technology, Surat, India

References

[1] Ernst R, Roland NI, Gordon AA. Rhodanine. *Org. Synth.* 1947, 27, 73.

[2] Nencki, M. Ueber die einwirkung der monochloressigsaure auf sulfocyansaure und ihre salze. *J. Prakt. Chem.* 1877, 16, 1–17.

[3] Tomasic T, Peterlin Masic L. Rhodanine as a scaffold in drug discovery: a critical review of its biological activities and mechanisms of target modulation. *Expert. Opin. Drug. Discov.* 2012, 7, 549–560.

[4] Baell JB. Observations on screening-based research and some concerning trends in the literature. *Future Med. Chem.* 2010, 2, 1529–1546.

[5] Mendgen T, Steuer C, Klein CD. Privileged Scaffolds or Promiscuous Binders: A Comparative Study on Rhodanines and Related Heterocycles in Medicinal Chemistry. *J. Med. Chem.* 2012, 55, 743–753.

[6] Allan FJ, Allan GG, Crank G, Jack J. The condensation of rhodanine and derivatives with benzaldehyde sulphonic acids. *Recl. Trav. Chim. Pays–Bas.* 1960, 79, 247–254.

[7] Allan FJ, Allan GG. The condensation of rhodanine and derivatives with 4-antipyrinaldehyde. *Can. J. Chem.* 1961, 39, 1397–1399.

[8] Allan FJ, Allan GG. Die Kondensation von Rhodanin und-derivaten mit einigen Indol–Aldehyden. *Monatsh. Chem.* 1963, 94, 569–573.

[9] Allan FJ, Allan GG. Isoxazolylmethylenerhodanines. *Recl. Trav. Chim. Pays–Bas.* 1964, 83, 1299–1300.

[10] Tanouchi T, Kawamura M, Ajima A, Mohri T, Hayashi M, Terashima H, Hirata F, Morimura T. US Pat 4464382A, 1984.

[11] Tanouchi T, Kawamura M, Ajima A, Mohri T, Hayashi M, Terashima H, Hirata F, Morimura T. US Pat 4831045A, 1989.

[12] Orchard MG, Neuss JD. WO Pat 0222612, 2002.

[13] Orchard MG. WO Pat 03070238, 2003.

[14] Inamori Y, Muro C, Tanaka R, Adachi A, Miyamotoand K, Tsujibo H. Phytogrowth-inhibitory activity of sulfur-containing compounds. I. Inhibitory activities of thiazolidine derivatives on plant growth. *Chem. Pharm. Bull.* 1992, 40, 2854–2856.

[15] Muro C, Yasuda M, Sakagami Y, Yamada T, Tsujibo H, Numata A, Inamorl Y. Inhibitory activities of rhodanine derivatives on plant growth. *Biasct. Biotech. Biochem.* 1996, 60, 1368–1371.

[16] Taniyama H, Yasui B, Takehara N, Uchida H. Studies on chemotherapeutics for micobacterium tuberculosis. XIX Synthesis and antibacterial activity of some 3-substituted rhodanines. *Yakugaku Zasshi.* 1959, 1465–1468.

[17] Frankov A, Kirillov MV, Sokolova TN, Skupskaya R, Kharitonovich AN, Chizhevskaya II. Synthesis and pharmacological properties of alkyl derivatives of 3-carboxyalkylrhodanine. *Khim. Farm. Zh.* 1985, 19, 943–946.

[18] Friebe WG, Krell HW, Woelle S, Wolff HP. WO Pat 0157006, 2001.

[19] Singh R, Ramesh UV, Goff D, Laidig G, Issakani SD, Huang J, Payan DG. WO Pat 2004043955, 2004.

[20] Tomasic T, Peterlin Masic L. Rhodanine as a privileged scaffold in drug discovery. *Curr. Med. Chem.* 2009, 16, 1596–1629.

[21] Villain–Guillot P, Gualtieri M, Bastide L, Roquet F, Martinez J, Amblard M, Pugniere M, Leonetti JP. Structure-activity relationships of phenyl-furanyl-rhodanines as in-

hibitors of RNA polymerase with antibacterial activity on biofilms. *J. Med. Chem.* 2007, 50, 4195–4204.

[22] Hardej D, Ashby CR, Khadtare NS, Kulkarni SS, Singh S, Talele TT. The synthesis of phenylalanine-derived C5-substituted rhodanines and their activity against selected methicillin-resistant Staphylococcus aureus (MRSA) strains. *Eur. J. Med. Chem.* 2010, 45, 5827–5832.

[23] Li W, Zheng CJ, Sun LP, Song MX, Wu Y, Li YJ, Liu Y, Piao HR. Novel arylhydrazone derivatives bearing a rhodanine moiety: synthesis and evaluation of their antibacterial activities. *Arch. Pharm. Res.* 2014, 37, 852–861.

[24] Zheng CJ, Song MX, Sun LP, Wu Y, Hong L, Piao HR. Synthesis and biological evaluation of 5-aryloxypyrazole derivatives bearing a rhodanine-3-aromatic acid as potential antimicrobial agents. *Bioorg. Med. Chem. Lett.* 2012, 22, 7024–7028.

[25] Xu LL, Zheng CJ, Sun LP, Miao J, Piao HR. Synthesis of novel 1,3-diaryl pyrazole derivatives bearing rhodanine-3-fatty acid moieties as potential antibacterial. *Eur. J. Med. Chem.* 2012, 48, 174–178.

[26] Miao J, Zheng C, Sun L, Song M, Xu L, Piao H. Synthesis and potential antibacterial activity of new rhodanine-3-acetic acid derivatives. *Med. Chem. Res.* 2013, 22, 4125–4132.

[27] Orchard MG, Neuss JC, Galley CM, Carr A, Porter DW, Smith P, Scopes DI, Haydon D, Vousden K, Stubberfield CR, Young K, Page M. Rhodanine-3-acetic acid derivatives as inhibitors of fungal protein mannosyl transferase 1 (PMT1). *Bioorg. Med. Chem. Lett.* 2004, 14, 3975–3978.

[28] Sortino M, Delgado P, Juárez S, Quiroga J, Abonía R, Insuasty B, Nogueras M, Rodero L, Garibotto FM, Enriz RD, Zacchino SA. Synthesis and antifungal activity of (Z)-5-arylidenerhodanines. *Bioorg. Med. Chem.* 2007, 15, 484–494.

[29] Chauhan K, Sharma M, Singh P, Kumar V, Shukla PK, Siddiqi MI, Chauhan PMS. Discovery of a new class of dithiocarbamates and rhodanine scaffold as potent antifungal agent: synthesis, biology and molecular docking. *Med. Chem. Commun.* 2012, 3, 1104–1110.

[30] Insuasty A, Ramírez J, Raimondi M, Echeverry C, Quiroga J, Abonia R, Nogueras M, Cobo J, Rodríguez MV, Zacchino SA, Insuasty B. Synthesis, antifungal and antitumor activity of novel (z)-5-hetarylmethylidene-1,3-thiazol-4-ones and (z)-5-ethylidene-1,3-thiazol-4-ones. *Molecules* 2013, 18, 5482–5497.

[31] Murugan R, Anbazhagan S, Lingeshwaran, Sriman Narayanan S. Synthesis and in vivo antidiabetic activity of novel dispiropyrrolidines through [3+2] cycloaddition reactions with thiazolidinedione and rhodanine derivatives. *Eur. J. Med. Chem.* 2009, 44, 3272–3279.

[32] Moorthy BT, Ravi S, Srivastava M, Chiruvella KK, Hemlal H, Joy O, Raghavan SC. Novel rhodanine derivatives induce growth inhibition followed by apoptosis. *Bioorg. Med. Chem. Lett.* 2010, 20, 6297–6301.

[33] Li W, Zhai X, Zhong Z, Li G, Pu Y, Gong P. Design, Synthesis and Evaluation of Novel Rhodanine-containing Sorafenib Analogs as Potential Antitumor Agents. *Arch. Pharm.* 2011, 344, 349–357.

[34] Liu J, Wu F, Chen L, Hu J, Zhao L, Chen C, Peng L. Evaluation of dihydropyrimidin-(2H)-one analogues and rhodanine derivatives as tyrosinase inhibitors. *Bioorg. Med. Chem. Lett.* 2011, 15, 2376–2379.

[35] Min G, Lee SK, Kim HN, Han YM, Lee RH, Jeong DG, Han DC, Kwon BM. Rhodanine-based PRL-3 inhibitors blocked the migration and invasion of metastatic cancer cells. *Bioorg. Med. Chem. Lett.* 2013, 23, 3769–3774.

[36] Rajamaki S, Innitzer A, Falciani C, Tintori C, Christ F, Witvrouw M, Debyser Z, Massa S, Botta M. Exploration of novel thiobarbituric acid-, rhodanine- and thiohydantoin-based HIV-1 integrase inhibitors. *Bioorg. Med. Chem. Lett.* 2009, 19, 3615–3618.

[37] Maga G, Falchi F, Radi M, Botta L, Casaluce G, Bernardini M, Irannejad H, Manetti F, Garbelli A, Samuele A, Zanoli S, Esté JA, Gonzalez E, Zucca E, Paolucci S, Baldanti F, De Rijck J, Debyser Z, Botta M. Toward the discovery of novel anti-HIV drugs. Second-generation inhibitors of the cellular ATPase DDX3 with improved anti-HIV activity: synthesis, structure–activity relationship analysis, cytotoxicity studies, and target validation. *ChemMedChem.* 2011, 6, 1371–1389.

[38] Jiang S, Tala SR, Lu H, Abo–Dya NE, Avan I, Gyanda K, Lu L, Katritzky AR, Debnath AK. Design, synthesis, and biological activity of novel 5-((arylfuran/1H-pyrrol-2-yl)methylene)-2-thioxo-3-(3-(trifluoromethyl)phenyl)thiazolidin-4-ones as HIV-1 Fusion Inhibitors Targeting gp41. *J. Med. Chem.* 2011, 54, 572–579.

[39] Talele TT, Arora P, Kulkarni SS, Patel MR, Singh S, Chudayeu M, Kaushik–Basu N. Structure-based virtual screening, synthesis and SAR of novel inhibitors of hepatitis C virus NS5B polymerase. *Bioorg. Med. Chem.* 2010, 18, 4630–4638.

[40] Patel BA, Krishnan R, Khadtare N, Gurukumar KR, Basu A, Arora P, Bhatt A, Patel MR, Dana D, Kumar S, Kaushik–Basu N, Talele TT. Design and synthesis of L- and D-phenylalanine derived rhodanines with novel C5-arylidenes as inhibitors of HCV NS5B polymerase. *Bioorg. Med. Chem.* 2013, 21, 3262–3271.

[41] Cutshall NS, O'Day C, Prezhdo M. Rhodanine derivatives as inhibitors of JSP-1. *Bioorg. Med. Chem. Lett.* 2005, 15, 3374–3379.

[42] Irvine MW, Patrick GL, Kewney J, Hastings SF, MacKenzie SJ. Rhodanine derivatives as novel inhibitors of PDE4. Bioorg. Med. Chem. Lett. 2008, 18, 2032–2037.

Chapter 3

Symmetrical Pyridinium-Phanes and –Diazacyclophanes — Promising Heterocyclic Scaffolds for the Development of Anti-Leishmanial Agents

Joaquín M. Campos, Verónica Gómez-Pérez, Santiago Castanys and Francisco Gamarro

Additional information is available at the end of the chapter

http://dx.doi.org/10.5772/61863

Abstract

There is an urgent need for better drugs for a more successful fight against leishmaniasis, one of the most important neglected diseases caused by the parasite *Leishmania*. We have recently synthesized several symmetrical pyridinium compounds belonging to two different series: bis-pyridinium and bis-quinolinium acyclic structures and bis-pyridinium diazacyclophanes derivatives. The first series of bis-pyridinium derivatives have been found to display activity against promastigotes and intracellular amastigotes of *Leishmania donovani* and *Leishmania major*, with EC_{50} values lower than 1 μM. The majority of compounds show a similar behavior in both *Leishmania* species, being slightly more active against intracellular amastigotes of L. *major*. The series of bis-pyridinium diazacyclophanes can be considered as rigid analogues of the previous bis-cationic ones. The activity of these compounds has also been evaluated against promastigotes and intracellular amastigotes of *L. donovani* and *L. major*. All the diazacyclophanes are more active against *L. major*, with EC_{50} values of between 1 and 17 μM in intracellular amastigotes, and in some cases they present a higher selectivity index than the reference anti-leishmanial drugs such as amphotericin B and miltefosine. In conclusion, these bis-quaternary compounds represent promising candidates as potential therapeutic agents against leishmaniasis.

Keywords: Pyridinium phanes, diazacyclophanes, leishmaniasis, *Leishmania* chemotherapy

1. Introduction

Leishmaniasis is a major group of neglected tropical diseases caused by the protozoan parasite *Leishmania*. Currently it affects 12 million people in 98 countries, and around 350 million people worldwide are at risk of infection [1]. Leishmaniasis is responsible for a variety of pathologies that have been classified in three main clinical manifestations including cutaneous (CL), mucocutaneous (MCL), and visceral (VL) leishmaniasis, ranging from self-healing cutaneous lesions to fatal visceral infection [2].

All *Leishmania* species are digenetic parasites that exist as both insect vector (promastigotes) and mammalian forms (intracellular amastigotes). The digenetic life cycle of *Leishmania* consists of flagellated, motile, extracellular promastigote form that proliferates in the midgut of phlebotomine sand fly family vectors, which infect mammalian host and transform into the non-motile, intracellular amastigote form that resides in phagolysosomes of macrophages and other reticuloendothelial cells.

Since an effective vaccine against leishmaniasis is not available, chemotherapy is at present the only effective way to treat all forms of the disease. The recommended first-line therapies for leishmaniasis include pentavalent antimonials such as sodium stibogluconate and meglumine antimoniate, amphotericin B (AmB), paromomycin, and miltefosine (Figure 1), all of which have different types of limitations including toxicity, price, efficacy, and emerging resistance [3], which emphasizes the importance of developing new drugs against leishmaniasis. Pentamidine [1,5-bis(4-amidinophenoxy)pentane] is an aromatic diamidine (Figure 1) widely used for the treatment of sleeping sickness caused by *Trypanosoma brucei* [4]. It was used as a second-line drug against VL in cases of antimony failure, but its use against leishmaniasis is now limited to the treatment of some forms of CL in South America [5]. Pentamidine acts at the mitochondrial level of the parasite by accumulating within the mitochondria and binding to DNA, thus interfering with the replication and transcription [6]. Novel diamidine derivatives with improved pharmacokinetic properties have been under development in recent years [7, 8].

New diamidine and choline-derivative dications have been developed recently in order to find new drugs with improved activity against leishmaniasis and lower toxicity [9–12] (Figure 2).

Chemistry is a science on which all the other sciences are based. An understanding of biology requires knowledge of chemistry. The majority of the leishmaniasis reviews are concentrating on the biology of the processes and very little on the chemistry. We would like to fill this gap and we will focus on the chemical structures that could be useful to the medicinal chemists working in this important area of research.

Here we present the anti-leishmanial activity of a set of symmetrical bis-pyridinium compounds with cyclic or acyclic structures. Both types of compounds can be named according to the IUPAC nomenclature for phanes, a method based on assembling names that describe component parts of a complex structure.

2. Symmetrical bis-pyridinium compounds

We have previously designed and synthesized a set of symmetrical bis-pyridinium compounds, which consist of a linker and two cationic heads which are 4-substituted pyridinium

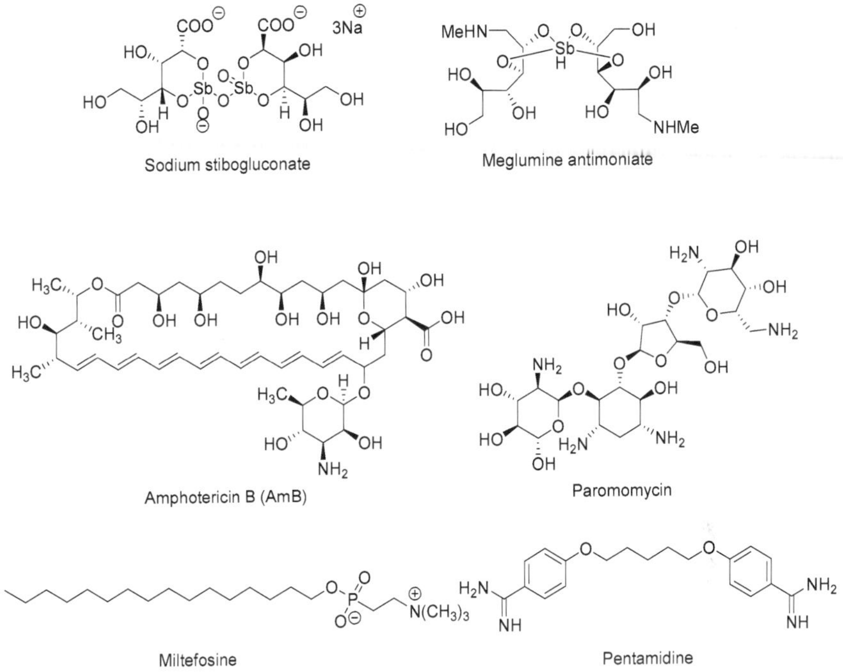

Figure 1. Anti-leishmanial drugs.

or quinolinium rings with cyclic or acyclic amino groups, as inhibitors of the human choline kinase (ChoK) (Table 1), the first enzyme in the CDP–choline pathway that synthesizes phosphatidylcholine, the major phospholipid in eukaryotic cell membranes. In humans, choline kinase exists as three isoforms (ChoKα1, α2, and β). Specific inhibition of ChoKα has been reported to selectively kill the tumor cells. Ten symmetrical bis-pyridinium and bis-quinolinium derivatives were tested for their ability to inhibit human ChoKα2, and **VGP-118** and **VGP-150** were identified as highly potent choline kinase inhibitors with EC_{50} values of 80 nM. Kinetic enzymatic assays indicated a mixed, predominantly competitive, inhibition mechanism for these compounds. These novel compounds showed strong anti-proliferative activity (EC_{50} of 1 μM) on the human breast cancer SKBR3 cell line [13].

In addition, these compounds can be considered as structural analogues of pentamidine in which the amidino moiety, which is protonated at physiological pH, has been replaced by a positively charged nitrogen atom as a pyridinium ring. In view of this structural resemblance and with the intention of identifying potential drugs against leishmaniasis, we analyzed the anti-leishmanial activity of these bis-pyridinium derivatives.

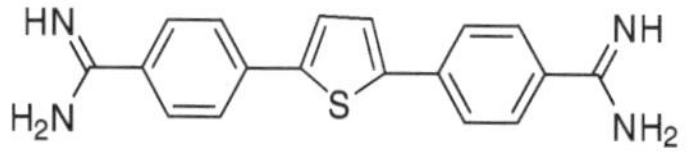

2,5-Bis-(4-amidinophenyl)thiophene
(ref. 9)

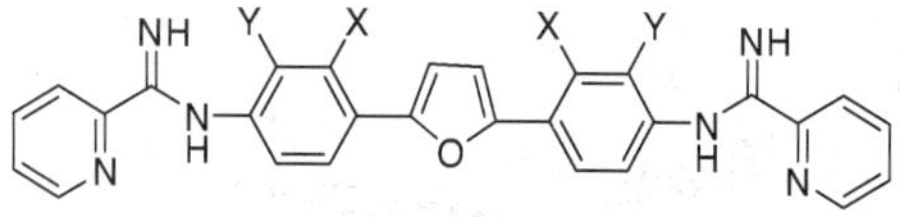

X = H; Y = OCH_3 **DB746**
X = OCH_3; Y = H **DB746**
(ref. 10)

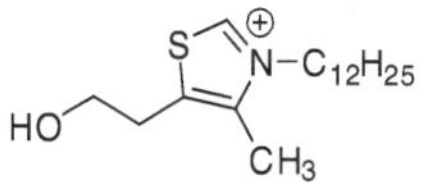

T1 (ref. 11)

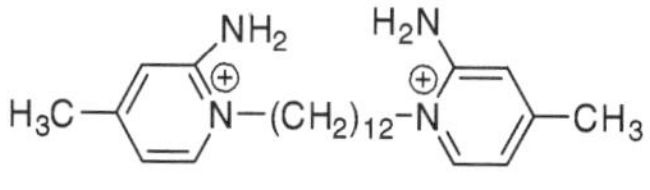

MS1 (ref 11)

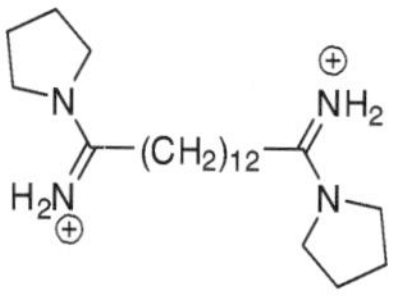

M38 (ref 11)

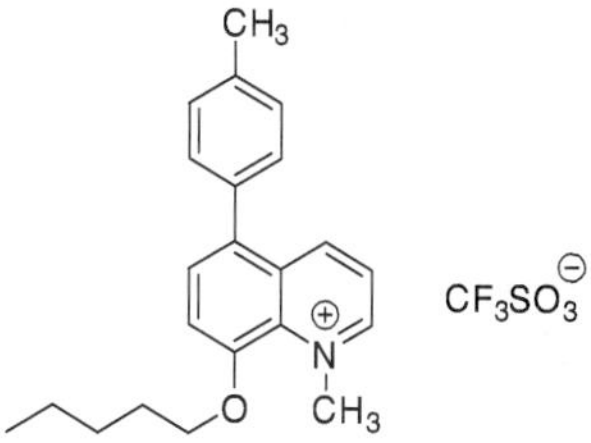

Quinolinium salt
(ref. 12)

Figure 2. Potent diamidine and charged derivatives with improved activity against leismaniasis.

2.1. Susceptibility analysis in *Leishmania* lines

The anti-leishmanial activity of the ten choline kinase inhibitors was evaluated against promastigotes and intracellular amastigotes of *Leishmania donovani* and *Leishmania major* in order to identify the potential hits for further optimization. The cytotoxic effect of these compounds was also investigated on the human monocytic cell line THP-1, the host cell used in the assay with intracellular amastigotes. Selectivity indexes (SI) were calculated as the ratio of the EC_{50} (the concentration of compound required to inhibit growth by 50%) for THP-1 to the EC_{50} for intracellular amastigotes. Table 2 shows the results, where miltefosine and AmB were used as the reference anti-leishmanial drugs. Most assayed compounds exhibit a specific high activity against promastigotes and intracellular amastigotes of *L. major*, with EC_{50} values between 0.09 and 0.42 μM in amastigotes, except for compounds **VGP-106** and **VGP-118** (EC_{50} 13.07 and 6.21 μM, respectively). With regard to *L. donovani*, all assayed compounds display EC_{50} values below 1 μM in promastigotes, except compound **VGP-138** (EC_{50} 2.11 μM). Although these values are slightly higher in intracellular amastigotes, they are similar to those for the anti-leishmanial drug miltefosine [14].

http://dx.doi.org/10.5772/61863

Our analysis of the effect on THP-1 cells showed that bis-pyridinium derivatives (**VGP-106**, **VGP-114**, **VGP-118**, **VGP-130**, **VGP-138**) are less cytotoxic than the bis-quinolinium counterparts (**VGP-146**, **VGP-150**, **VGP-162**, **VGP-174**, **VGP-182**), with a higher SI than miltefosine (Table 2).

Compound	Het	Linker
VGP-106	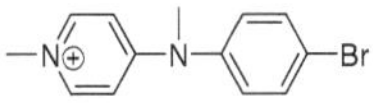	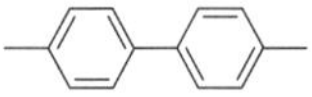
VGP-114	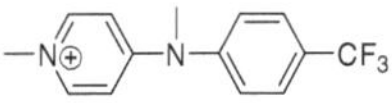	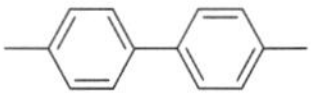
VGP-118	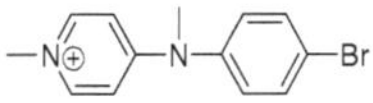	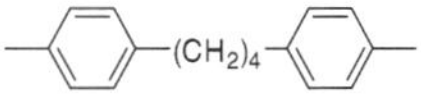
VGP-130	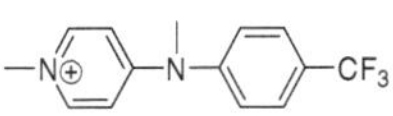	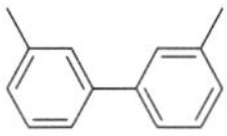
VGP-138	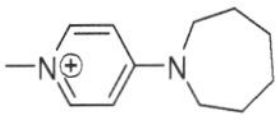	
VGP-146	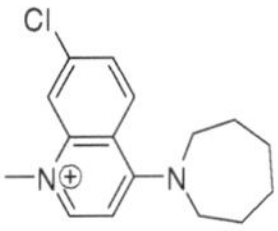	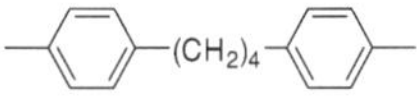
VGP-150	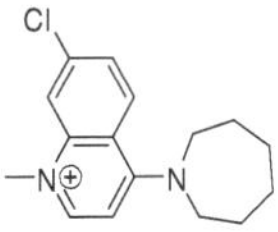	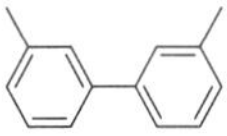
VGP-162	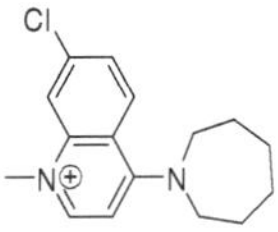	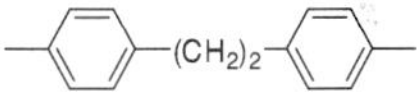
VGP-174	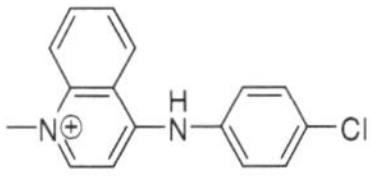	
VGP-182	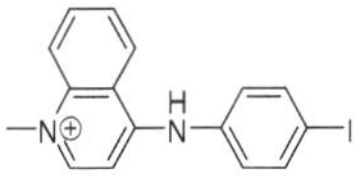	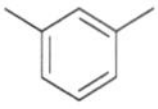

Table 1. Structure of the bis-cationic compounds

	EC_{50} promastigotes (μM)		EC_{50} amastigotes (μM) [SI][b]		THP-1 toxicity EC_{50} (μM)
Compound	*L. major*	*L. donovani*	*L. major*	*L. major*	
VGP-106	21.55 ± 3.72	0.36 ± 0.09	13.07 ± 6.30 [15.8]	0.86 ± 0.46 [240.2]	206.54 ± 9.89
VGP-114	0.47 ± 0.04	0.61 ± 0.09	0.10 ± 0.03 [1000.6]	0.85 ± 0.04 [117.7]	100.06 ± 8.57
VGP-118	29.15 ± 5.73	0.65 ± 0.19	6.21 ± 1.02 [2.4]	0.18 ± 0.03 [85.3]	15.35 ± 3.99
VGP-130	0.50 ± 0.07	0.73 ± 0.11	0.09 ± 0.02 [903.7]	2.02 ± 0.05 [40.3]	81.34 ± 10.65
VGP-138	0.74 ± 0.19	2.11 ± 0.48	0.30 ± 0.16 [586.8]	4.01 ± 0.43 [43.9]	176.05 ± 20.75
VGP-146	0.21 ± 0.06	0.33 ± 0.07	0.10 ± 0.04 [156.1]	0.42 ± 0.01 [37.2]	15.61 ± 3.26
VGP-150	0.36 ± 0.11	0.77 ± 0.04	0.09 ± 0.03 [267]	0.55 ± 0.16 [43.7]	24.03 ± 5.42
VGP-162	0.40 ± 0.08	0.35 ± 0.02	0.37 ± 0.03 [29.6]	1.00 ± 0.08 [11.0]	10.97 ± 2.41
VGP-174	1.70 ± 0.01	0.34 ± 0.03	0.41 ± 0.05 [6.1]	0.86 ± 0.03 [2.8]	2.47 ± 0.05
VGP-182	2.51 ± 0.01	0.92 ± 0.2	0.42 ± 0.12 [11.2]	0.52 ± 0.12 [9.1]	4.71 ± 0.23
AmB	0.32 ± 0.02	0.21 ± 0.01	0.24 ± 0.01 [59.7]	0.28 ± 0.13 [51.1]	14.32 ± 4.10
Miltefosine	16.65 ± 1.23	6.60 ± 1.57	10.61 ± 0.89 [2.5]	0.88 ± 0.14 [30.5]	26.86 ± 3.08

[a]Parasites were grown for 72 h at 28 °C (promastigotes) or 37 °C (intracellular amastigotes) in the presence of increasing concentrations of compounds. THP-1 cells were grown for 72 h at 37 °C, in the presence of increasing concentrations of compounds. Promastigotes and THP-1 viability was determined using an MTT-based assay. Number of intracellular amastigotes was determined by nuclear staining. AmB and miltefosine were used as standard anti-leishmanial agents. Data are means ± SD of three independent determinations.

[b]Selectivity indexes [SI] were calculated by dividing the EC_{50} THP-1 by that for intracellular amastigotes. Compound **VGP-106** (grey color) was selected for further studies of the mechanism of action.

Table 2. Anti-leishmanial activity and toxicity in THP-1 cells of symmetrical bis-pyridinium compounds.[a]

Compound **VGP-106** was identified as a representative compound that displayed a potent activity against *L. donovani* intracellular amastigotes. As the least cytotoxic of the set of compounds assayed for THP-1 cells, it was selected to further elucidate their mechanism of action in this protozoan parasite [14].

2.2. Drug susceptibility assay of *L. donovani* lines overexpressing CEK or EK

Considering that the *Leishmania* genome includes two homologous enzymes of human ChoK, namely, choline/ethanolamine kinase (CEK) and ethanolamine kinase (EK), we decided to study whether there is a correlation between their ChoK inhibitory activity and anti-leishmanial activity. These proteins can be overexpressed in *L. Donovani* promastigotes by transfecting the parasites with a plasmid encoding the *Leishmania* CEK (pXG-CEK) or EK (pXG-EK) genes [14]. The susceptibility of transfected parasites to compound **VGP-106** was determined in both promastigotes and intracellular amastigotes. As can be seen from Table 3, there are no significant differences between the EC_{50} values of parasites overexpressing CEK or EK enzymes compared to control parasites. These results suggest that the mechanism of action of this compound in *Leishmania* is independent of the aforementioned enzymes [14]. If this were not the case, overexpression of these enzymes would have resulted in an increase in the EC_{50} value.

	EC_{50} (μM)	
Plasmid	**Promastigotes**	**Amastigotes**
pXG	0.36 ± 0.09	0.45 ± 0.03
pXG-CEK	0.36 ± 0.09	0.42 ± 0.05
pXG-EK	0.43 ± 0.05	0.35 ± 0.03

[a]Control (pXG) and transfected (pXG-CEK and pXG-EK) parasites were grown for 72 h at 28 °C (promastigotes) or 37 ºC (intracellular amastigotes) in the presence of increasing concentrations of compounds. Data are means ± SD of three independent determinations.

Table 3. Susceptibility to VGP-106 of *L. donovani* lines overexpressing CEK or EK.[a]

3. Symmetrical bis-pyridinium diazacyclophanes

Rigidification is a commonly used strategy to increase the activity of a drug or to reduce its side effects. A cyclophane is a hydrocarbon consisting of an aromatic unit (typically a benzene ring) and an aliphatic chain that forms a bridge between two non-adjacent positions of the aromatic ring.

We have synthesized a new family of symmetrical bis-pyridinium diazacyclophanes designed as cyclic analogues of previously reported acyclic bis-pyridinium derivatives, by cyclization through the exocyclic nitrogen atoms at position 4 of the pyridinium moiety via linker 2, which leads to the diazacyclophane targets (Figure 3) [15]. These compounds have been evaluated against *L. major* and *L. donavani*.

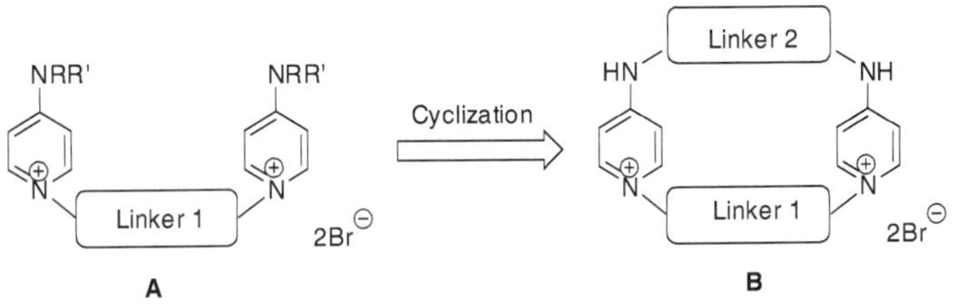

Figure 3. Structural variation that leads to symmetrical bis-pyridinium diazacyclophanes (**B**) from symmetrical acyclic bis-pyridinium derivatives (**A**).

This new compounds are symmetrical bis-pyridinium derivatives which differ from each other in the upper and lower spacers. Four different spacers were used: two are phenyl-*p*-diyl-methylene and phenyl-*m*-diylmethylene linkers, and the other two are aliphatic, such as the

1,5-pentanediyl and 3-oxa-1,5-pentanediyl moieties. At least one of the two spacers in every cyclophane is an aliphatic linker (Table 4).

The final compounds were synthesized according to Scheme 1. Dipyridines **1** and **2** were prepared from commercially available diamines and 4-bromopyridine in the presence of phenol under argon atmosphere, as previously described [11]. A reaction involving phenol as proton donor, solvating agent, and source of phenoxide ion is envisaged, as outlined in Scheme 1 [16]. As a reaction medium, phenol reduces both the reaction time and temperature of the halogen-replacement reactions.

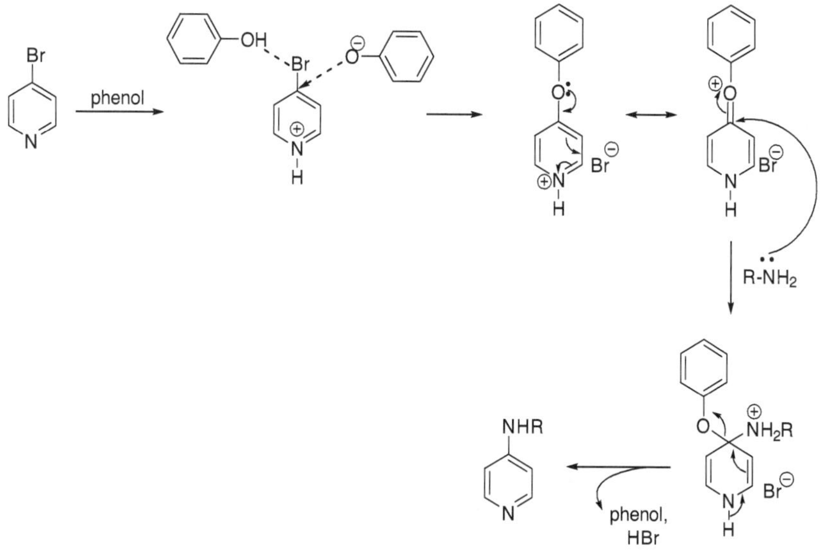

Sheme 1. As a reaction medium, phenol reduces reaction time and temperature of halogen-replacement reactions, by acting as proton donor, solvating agent, and source of phenoxide ion.

The novel dipyridines (**3** and **4**) were prepared from commercially available pentane-1,5-diamine and bis-2-(aminoethyl)ether, and following the same synthetic protocol previously reported [11].

Cyclophanes were obtained by cyclization of dipyridines **1-4** and the dibromide derivatives in acetonitrile, according to our reported procedures [11]. The reaction was carried out by adding 4 mM solution of the dibromide drop by drop to the dipyridine in acetonitrile at the reflux temperature of the mixture for a period of 10–12 days, which favors the cyclization step and avoids the intermolecular reaction [17]. In order to shorten the reaction time, microwave was used. Thus, dipyridine and dibromide derivatives in acetonitrile were microwave-irradiated at 140 ºC for 20 min. Under these conditions, similar yields were obtained as compared to standard heating at the boiling point of the solvent (acetonitrile). Similarly bis-

http://dx.doi.org/10.5772/61863

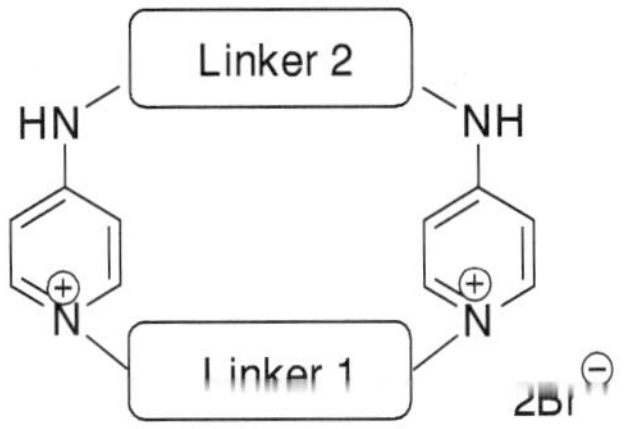

Compound	Linker 1	Linker 2
VGP-222	$-(CH_2)_5-$	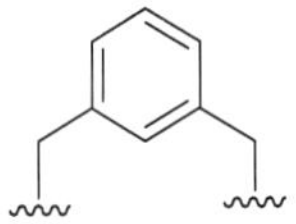
VGP-234	$-(CH_2)_2\text{-}O-(CH_2)_2-$	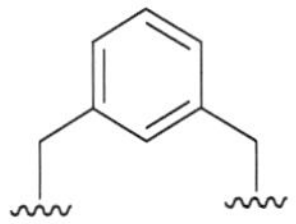
VGP-310	$-(CH_2)_5-$	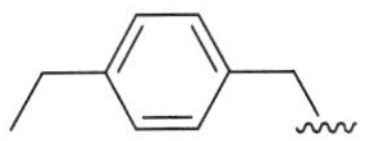
VGP-312	$-(CH_2)_2\text{-}O-(CH_2)_2-$	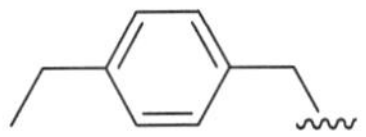
VGP-318	$-(CH_2)_5-$	$-(CH_2)_5-$
VGP-328	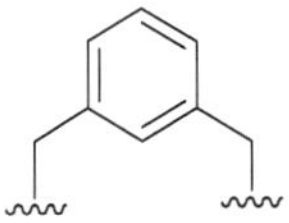	$-(CH_2)_2\text{-}O-(CH_2)_2-$
VGP-334	$-(CH_2)_5-$	$-(CH_2)_2\text{-}O-(CH_2)_2-$
VGP-340	$-(CH_2)_2\text{-}O-(CH_2)_2-$	$-(CH_2)_2\text{-}O-(CH_2)_2-$
VGP-352	$-(CH_2)_2\text{-}O-(CH_2)_2-$	$-(CH_2)_5-$

Table 4. Structures of the symmetrical bis-pyridinium diazacyclophanes.

quinolinium cyclophanes [18–20] needed to be purified by tedious reverse-phase preparative HPLC because conventional purification methods failed to give analytically pure samples for biological testing, despite having been obtained under high-dilution conditions (1–2 mM). In our case, this represents a great advantage for the accessibility of such an interesting class of compounds (Scheme 2).

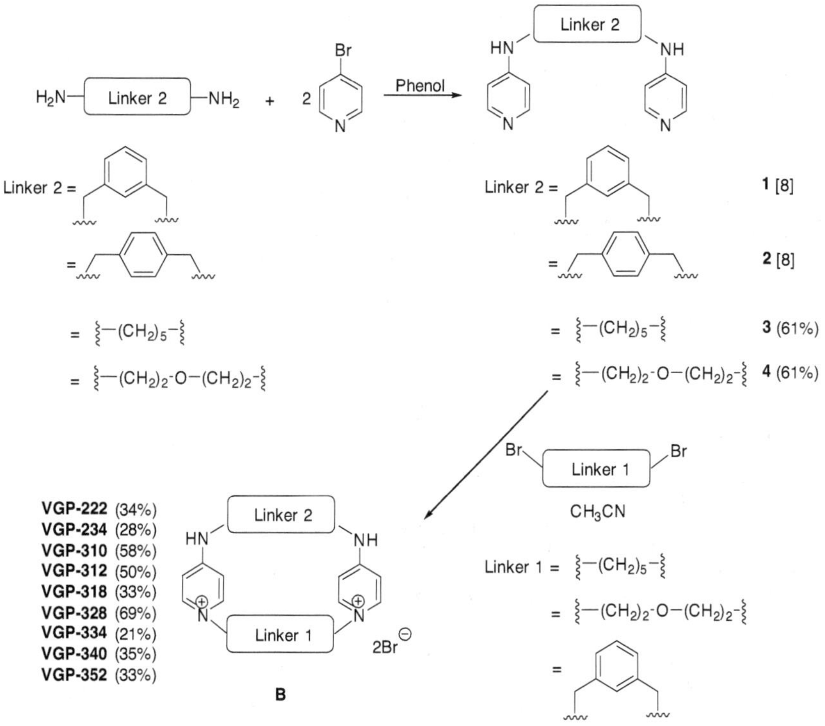

Sheme 2. General synthesis of the symmetrical bis-pyridinium diazacyclophanes.

3.1. Anti-leishmanial activity

The final nine cyclophanes were tested as anti-leishmanial agents against promastigotes and intracellular amastigotes of *L. donovani* and *L. major* [15]. The results are shown in Table 5, where miltefosine and AmB were used as reference drugs.

All assayed compounds exhibit activity against promastigotes and intracellular amastigotes of *L. major* and *L. donovani*, being more active in *L. major*, with EC_{50} values lying in the range 1 and 17 μM in amastigotes. Compounds **VGP-310, VGP-318, VGP-334, VGP-340,** and **VGP-352** display EC_{50} values below 1 μM against promastigotes of *L. major*, an activity 100-fold higher than that obtained in promastigotes of *L. donovani*. However, the differences in activity decrease in the amastigote forms, because some of these compounds are less active in amastigotes than in promastigotes of *L. major* and more active in amastigotes than in promastigotes of *L. donovani*.

	EC_{50} promastigotes (μM)		EC_{50} amastigotes (μM) [SI][b]		THP-1 toxicity EC_{50} (μM)
Compound	*L. major*	*L. donovani*	*L. major*	*L. donovani*	
VGP-222	16.84 ± 1.20	51.97 ± 1.97	5.94 ± 0.93 [32.3]	13.53 ± 1.40 [14.2]	191.90 ± 8.12
VGP-234	5.97 ± 0.35	33.77 ± 4.68	8.67 ± 1.04 [22.5]	8.92 ± 1.96 [10.3]	195.17 ± 6.41
VGP-310	0.17 ± 0.01	26.41± 1.28	0.97 ± 0.27 [170.2]	38.33 ± 1.74 [4.3]	165.06 ± 21.29
VGP-312	26.48 ± 2.44	76.87 ± 9.59	17.15 ± 1.50 [12.9]	63.67 ± 5.21 [3.5]	221.89 ± 8.27
VGP-318	0.07 ± 0.01	10.64 ± 1.03	1.26 ± 0.30 [122.3]	7.62 ± 0.16 [20.2]	154.07 ± 5.95
VGP-328	2.87 ± 0.36	76.27 ± 4.96	1.61 ± 0.35 [120.8]	21.25 ± 2.03 [9.2]	194.41 ± 2.95
VGP-334	0.26 ± 0.02	31.47 ± 2.53	2.59 ± 0.23 [62.7]	33.19 ± 0.57 [4.9]	162.44 ± 6.07
VGP-340	0.19 ± 0.01	23.43 ± 0.57	2.24 ± 0.35 [57.2]	20.72 ± 1.07 [6.2]	128.22 ± 9.78
VGP-352	0.26 ± 0.01	31.41 ± 3.02	2.18 ± 0.05 [98.5]	12.95 ± 1.86 [16.6]	214.65 ± 13.80
AmB	0.32 ± 0.02	0.21 ± 0.01	0.24 ± 0.01 [59.7]	0.28 ± 0.13 [51.1]	14.32 ± 4.10
Miltefosine	16.65 ± 1.23	6.60 ± 1.57	10.61 ± 0.89 [2.5]	0.88 ± 0.14 [30.5]	26.86 ± 3.08

[a]Parasites were grown for 72 h at 28 °C (promastigotes) or 37 ºC (intracellular amastigotes) in the presence of increasing concentrations of compounds. THP-1 cells were grown for 72 h at 37 ºC, in the presence of increasing concentrations of compounds. Promastigote and THP-1 viability was determined using an MTT-based assay. Number of intracellular amastigotes was determined by nuclear staining. AmB, and miltefosine were used as standard anti-leishmanial agents. Data are means ± SD of three independent determinations.

[b]Selectivity indexes [SI] were calculated by dividing the EC_{50} THP-1 by that for intracellular amastigotes. Compound **VGP-318** (grey color) was selected for further studies of the mechanism of action.

Table 5. Anti-leishmanial activity and toxicity in THP-1 cells of symmetrical bis-pyridinium diazacyclophanes[a].

In general, from a structural point of view, compounds with two aliphatic linkers show better activity against promastigotes of *L. major* than compounds with an aromatic linker. However, the presence of an aromatic spacer increases the activity in intracellular amastigotes relative to the activity in promastigotes, except for **VGP-310**. This could be due to the higher lipophilicity of these structures that allowed a better penetration into THP-1 cells. Nevertheless, most compounds displayed higher activity in intracellular amastigotes than in promastigotes of *L. donovani*. Regarding the aliphatic linker, the presence of an oxygen atom in the linker did not involve significant differences in the activity. All diazacyclophanes exhibited very low toxicity against THP-1 cells (EC_{50} values between 128 and 220 μM) and some of them evince a higher selectivity index than the reference compounds.

Compound **VGP-318** was chosen as a representative compound to further investigate the mechanism of action of this new family of compounds [15]. This compound shows promising activity against intracellular amastigotes of *L. major* (EC_{50} 1.3 ± 0.3 μM), with a selectivity index (122) higher than those of AmB (51) and miltefosine (30). It is also the most active diazacyclophane derivative against intracellular amastigotes of *L. donovani* (EC_{50} 7.6 ± 0.2 μM).

3.2. Drug susceptibility assay of *L. donovani* lines overexpressing CEK or EK

As we have previously published that other bis-pyridinium diazacyclophanes were ChoK inhibitors and active anti-proliferative drugs [11], we performed a sensitivity test for **VGP-318** in promastigotes and intracellular amastigotes of *L. donovani* overexpressing the *Leishmania* enzymes CEK or EK. The sensitivity for **VGP-318** is similar in both promastigotes and intracellular amastigotes overexpressing CEK and EK versus control parasites (Table 6). This result suggests that the anti-leishmanial activity of these compounds is not related to the CEK and EK enzymes [15].

	EC_{50} (μM)	
Plasmid	**Promastigotes**	**Amastigotes**
pXG	13.50 ± 0.32	8.84 ± 0.18
pXG-CEK	11.51 ± 0.52	12.70 ± 2.03
pXG-EK	12.04 ± 0.42	10.54 ± 1.42

[a]Control (pXG) and transfected (pXG-CEK and pXG-EK) parasites were grown for 72 h at 28 °C (promastigotes) or 37 ºC (intracellular amastigotes) in the presence of increasing concentrations of compound. Data are means ± SD of three independent determinations.

Table 6. Susceptibility to VGP-318 of *L. donovani* lines overexpressing CEK or EK.[a]

3.3. Effect of VGP-318 on *Leishmania* metabolism

In order to investigate the anti-leishmanial mechanism of action of compound **VGP-318**, we focused the studies on the energetic metabolism of *Leishmania* promastigotes [15]. First, the effect of the compound on intracellular ATP levels was analyzed by the bioluminescence assay, which generates a luminescent signal proportional to the amount of ATP. In *L. major*, this assay showed a rapid decrease in the intracellular ATP levels which depends on the compound concentration (Figure 4A). However, no effect was observed on *L. donovani* after incubation with 30 μM for 3 h (Figure 4B). The decrease in the ATP levels may be caused mainly by an effect on the ATP synthesis or a release of the intracellular ATP due to the permeabilization of plasma membrane. However, under conditions that decrease 95% of the ATP relative to control (30 μM for 3 h), no sign of plasma membrane alteration was observed (Figure 5), showing that the drop of free intracellular ATP is not due to disruption of the plasma membrane and suggesting that this may be due to a defect in the ATP synthesis.

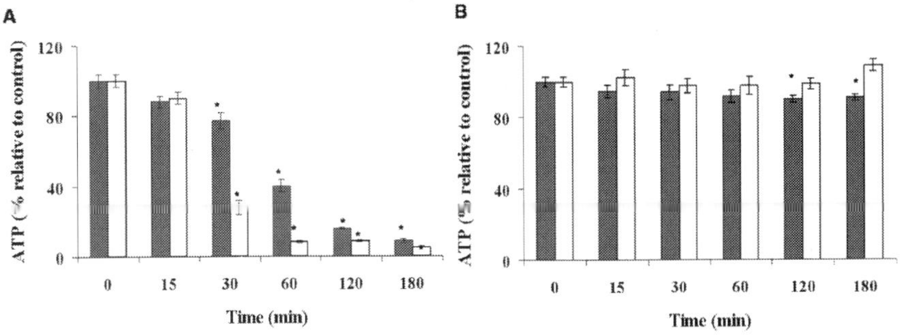

Figure 4. Effect of **VGP-318** on ATP levels in *Leishmania* promastigotes. Changes in intracellular ATP levels in *L. major* (A) or *L. Donovani* (B) promastigotes treated with 0.2 (black bar) or 30 μM (white bar) of compound **VGP-318** were determined using the bioluminescence assay. Data are means ± SD of three independent experiments. Significant differences were determined using Student's *t*-test (*p<0.01).

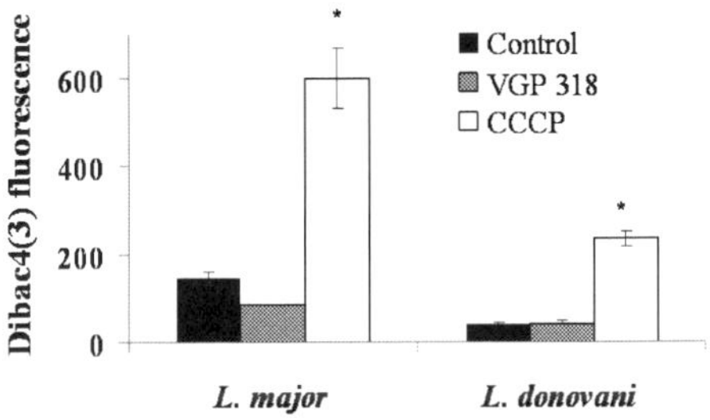

Figure 5. Compound **VGP-318** does not alter the plasma membrane potential in *Leishmania* lines. Promastigotes were incubated with 30 μM of compound **VGP-318** for 3 h and then treated with 2 μM of the specific plasma membrane potential probe DIBAC4(3) for 10 min at 28 ºC. Untreated parasites were used as control, and treatment with 10 μM CCCP was used as 100% depolarization of the plasma membrane potential. Data are means ± SD of three independent experiments. Significant differences were determined using Student's *t*-test (*p< 0.01).

In *Leishmania*, the ATP is mainly synthesized by mitochondrial oxidative phosphorylation [21]. The anti-parasitic activity of many drugs, such as pentamidine and miltefosine, is mediated by an alteration of the mitochondrial membrane potential ($\Delta\Psi_m$) [22, 23]. To determine whether the intracellular ATP decay was associated with an effect of **VGP-318** on the mitochondria, the variation of its electrochemical potential was monitored in parasites incubated with **VGP-318** using the JC-1 fluorescent marker. *L. major* parasites incubated for 1 h with 0.2 μM of **VGP-318** (conditions where there was 50% decay in the ATP levels) showed a significant decrease in JC-1 ratio compared with untreated parasites (Figure 6), evidencing a depolarization of the mitochondrial potential. However, in *L. donovani* promastigotes non-significant depolarization was observed after treatment with 30 μM of **VGP-318** for 3 h (Figure 6). The depolarization of the $\Delta\Psi_m$ in *L. major* promastigotes suggests that this compound may cause damage in the mitochondria, leading to a fall in the intracellular ATP levels and the death of parasites.

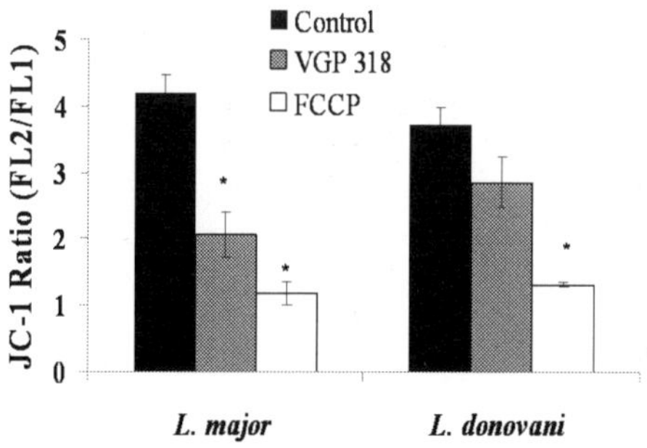

Figure 6. Effect of **VGP-318** on the $\Delta\Psi_m$ of *Leishmania* promastigotes. Promastigotes were treated with 0.2 μM (*L. major*) or 30 μM (*L. donovani*) of compound **VGP-318** for 3 h and then incubated with 5 μM JC-1 for 10 min for the ΔΨm determination. The FL2/FL1 fluorescence ratio was measured by flow cytometry analysis. Untreated parasites were used as control, and treatment with 10 μM FCCP for 10 min was used as full depolarization controls. Data are means ± SD of three independent experiments. Significant differences were determined using Student's *t*-test (*p<0.01).

The lack of effect of compound **VGP-318** in *L. donovani* may be explained by the lower activity of this compound; **VGP-318** is 100-fold less active in promastigote forms of *L. donovani* than in *L. major*. Additionally, **VGP-318** induces a slight depolarization of the $\Delta\Psi_m$ in *L. Donovani* promastigotes, suggesting that a longer incubation time is necessary to produce significant mitochondrial damage leading to failure of ATP synthesis. Compound **VGP-318** has been highlighted very recently [24].

4. Conclusions

In the search of new drugs against leishmaniasis, we have synthesized and evaluated two set of symmetrical bis-pyridinium derivatives: (i) bis-pyridinium and bis-quinolinium acyclic structures which contain a linker and 4-substituted cyclic or acyclic amino groups in the two cationic heads and (ii) bis-pyridinium diazacyclophanes that are rigid derivatives with an upper spacer which joins the two exocyclic amino groups and a lower spacer joining the two positively charged nitrogen atoms. Restriction of conformational flexibility could be an important consideration for the design of anti-leishmanial agents. Global constraint was obtained by backbone cyclization in a tail-to-tail fashion. This popular tactic in medicinal chemistry remains in some extent empirical, but has met successes, mainly for the elaboration of working or preliminary pharmacophores.

All these bis-pyridinium salts show activity against promastigotes and intracellular amastigotes of the protozoan parasites *L. donovani* and *L. major* [14, 15]. Most acyclic compounds show a similar behavior in both species, being slightly more active against *L. major* amastigotes. All the cyclophanes are more active against promastigotes and amastigotes of *L. major* than *L. donovani*, although with a lower potency than the acyclic derivatives. However, in contrast to the variable toxicity of the acyclic compounds [14], all cyclophanes exhibit very low toxicity against mammalian cells THP-1 and some of them evince a higher safety margin than well-known anti-leishmanial drugs such as AmB and miltefosine [15].

Although we have studied certain aspects of the mechanism of action of these compounds [14, 15], it has not been determined any key target on which they are operating, which would be decisive for the rational design of new structures. Future work should be directed to carry out studies to elucidate the metabolism, pharmacokinetics, and mechanism of action of these compounds. On the other hand, it would be interesting to conduct a screening of a large number of symmetrical bis-pyridinium compounds that allows us to study structure–activity relationships. In any case, additional experiments are necessary for evaluating the toxicity and potency of these compounds by *in vivo* assays.

Note

Some parts of this chapter have been previously published in references [14, 15].

Acknowledgements

This work was supported by the Spanish Grants SAF2012-34267 (to F.G.), SAF2011-28102 (to S.C.), the Plan Andaluz de Investigación (Proyecto de Excelencia CTS-7282), by FEDER funds from the EU to F.G. and S.C. and by an FPU fellowship (AP2009-3910) from the Ministerio de Educación (to V.G.P).

Author details

Joaquín M. Campos[1,2*], Verónica Gómez-Pérez[3], Santiago Castanys[3] and Francisco Gamarro[3]

*Address all correspondence to: jmcampos@ugr.es

1 Departamento de Química Farmacéutica y Orgánica, Facultad de Farmacia. Universidad de Granada, c/ Campus de Cartuja s/n, Granada, Spain

2 Instituto Biosanitario de Granada (ibs.GRANADA), SAS-Universidad de Granada, Granada, Spain

3 Instituto de Parasitología y Biomedicina "López-Neyra", IPBLN-CSIC, Parque Tecnológico de Ciencias de la Salud, Avda. del Conocimiento s/n, Armilla, Granada, Spain

References

[1] Alvar J, Vélez ID, Bern C, Herrero M, Desjeux P, Cano J, Jannin J, den Boer M. WHO Leishmaniasis Control Team, Leishmaniasis worldwide and global estimates of its incidence. Plos One 2012;7e35671. Doi: 10.1371/journal.pone.0035671.

[2] Leishmaniais. http://www.who.int/leishmaniasis/en/ (accessed 24 August 2015).

[3] Seifert K. Structures, targets and recent approaches in anti-leishmanial drug discovery and development. Open Medicinal Chemistry Journal 2011;5, 31–39. Doi: 10.2174/1874104501105010031.

[4] Fairlamb AH, Chemotherapy of human African trypanosomiasis: current and future prospects. Trends in Parasitology 2003;19, 488–494. Doi: 10.1016/j.pt.2003.09.002.

[5] van der Meide WF, Sabajo LOA, Jensema AJ, Peekel I, Faber WR, Schallig HD, Fat RF. Evaluation of treatment with pentamidine for cutaneous leishmaniasis in Suriname. International Journal of Dermatology 2009;48, 52–58. Doi: 10.1111/j.1365-4632.2009.03883.x.

[6] Sun T, Zhang Y. Pentamidine binds to tRNA through non-specific hydrophobic interactions and inhibits aminoacylation and translation. Nucleic Acids Research 2008;36 1654–1664. Doi: 10.1093/nar/gkm1180.

[7] Patric DA, Bakunov SA, Bakunova SM, Kumar EV, Chen H, Jones SK, Wenzler T, Barzcz T, Werbovetz KA, Brun R, Tidwell RR. Synthesis and antiprotozoal activities of dicationic bis(phenoxymethyl)benzenes, bis(phenoxymethyl)-naphthalenes, and bis(benzyloxy)-naphthalenes. European Journal of Medicinal Chemistry 2009;44, 3543–3551. Doi: 10.1016/j.ejmech.2009.03.014.

[8] Patric DA, Bakunov SA, Bakunova SM, Jones SK, Wenzler T, Barzcz T, Kumar A, Boykin DW, Werbovetz KA, Brun R, Tidwell RR. Synthesis and antiprotozoal activities of benzyl phenyl ether diamidine derivatives. European Journal of Medicinal Chemistry 2013;67, 310–324. Doi: 10.1016/j.ejmech.2013.06.033.

[9] Brendle JJ, Outlaw A, Kumar A, Boykin DW, Patrick DA, Tidwell RR, Werbovetz KA. Antileishmanial activities of several classes of aromatic dications. Antimicrobial Agents and Chemotherapy 2002;46797–46807. Doi:10.1128/AAC.46.3.797-807.2002.

[10] Rosypal AC, Werbovetz KA, Salem M, Stephens CE, Kumar A, Boykin DW, Hall JE, Tidwell RR. Inhibition by Dications of *in vitro* growth of Leishmania major and Leishmania tropica: causative agents of old world cutaneous leishmaniasis. Journal of Parasitology 2008;94, 743–749. Doi: 10.1645/GE-1387.1.

[11] Ibrahim HM, Al-Salabi MI, El Sabbagh N, Quashie NB, Alkhaldi AA, Escale R, Smith TK, Vial HJ, de Koning HP. Symmetrical choline-derived dications display strong anti-kinetoplastid activity. Journal of Antimicrobial Chemotherapy. 2011;66, 111–125. Doi: 10.1093/jac/dkq401.

[12] Bringmann G, Thomale K, Bischof S, Schneider C, Schultheis M, Schwarz T, Moll H, Schurigt U. A novel Leishmania major Amastigote Assay in 96-well format for rapid drug screening and its use for the discovery and evaluation of a new class of Leishmanicidal quinolinium salts. Antimicrobial Agents and Chemotherapy 2013;57, 3003–3011. Doi:10.1128/AAC.02201-12.

[13] Gómez-Pérez V, McSorley T, See Too WC, Konrad M, Campos JM. Novel 4-Amino Bis-pyridinium and bis-quinolinium derivatives as choline kinase inhibitors with antiproliferative activity against the human breast cancer SKBR-3 cell line. ChemMedChem 2012;7, 663–669. Doi: 10.1002/cmdc.201100505.

[14] Gómez-Pérez V, Manzano JI, García-Hernández R, Castanys S, Campos Rosa JM, Gamarro F. 4-Amino bis-pyridinium derivatives as novel anti-leishmanial agents. Antimicrobial Agents and Chemotherapy 2014;58, 4103–4112. Doi: 10.1128/AAC.02481-13.

[15] Gómez-Pérez V, Manzano JI, García-Hernández R, Castanys S, Gamarro F, Campos Rosa JM. Design, synthesis and anti-leishmanial activity of novel symmetrical bispyridinium cyclophanes. European Journal of Medicinal Chemistry 2015;89, 362–369. Doi: 10.1016/j.ejmech.2014.10.040.

[16] Surrey AR, Cutler RK. The role of phenol in the reaction of 4,7-dichloroquinoline with novol diamine. Journal of the American Chemical Society 1951;73, 2623–2626. Doi:10.1021/ja01150a060.

[17] Conejo-García A, Campos JM, Sánchez-Martín RM, Gallo MA, Espinosa A. Bis-pyridinium cyclophanes: novel templates for human choline kinase inhibitors. Journal of Medicinal Chemistry 2003;46 3754–3757. Doi:10.1021/jm030792i.

[18] Campos Rosa J, Galanakis D, Ganellin CR, Dunn PM, Jenkinson DH. Bis-Quinolinium Cyclophanes: 6,10-Diaza-3(1,3),8(1,4)-dibenzena-1,5(1,4)-diquinolinacyclodecaphane (UCL 1684), the first nanomolar, non-peptidic blocker of the apamin-sensitive $Ca^{2\pm}$-activated $K^{\pm}$ Channel. Journal of Medicinal Chemistry 1998;41, 2–5. Doi: 10.1021/jm970571a.

[19] Campos Rosa J, Galanakis D, Piergentili A, Bhandari K, Ganellin CR, Dunn PM, Jenkinson DH. Synthesis, molecular modeling, and pharmacological testing of bis-quinolinium cyclophanes: Potent, non-peptidic blockers of the apamin-sensitive Ca^{2+}-activated K^{+}channel. Journal of Medicinal Chemistry 2000;43, 420–431. Doi:10.1021/jm9902537.

[20] Chen JQ, Galanakis D, Ganellin CR, Dunn PM, Jenkinson DH. Bis-quinolinium cyclophanes: 8,14-diaza-1,7(1,4)-diquinolinacyclotetradecaphane (UCL 1848), a highly potent and selective, non-peptidic blocker of the apamin-sensitive Ca^{2+}-activated K^{+} channel. Journal of Medicinal Chemistry 2000;43, 3478–3481. Doi: 10.1021/jm000904v

[21] Tielens AG, van Hellemond JJ. Surprising variety in energy metabolism within Trypanosomatidae. Trends Parasitology 2009;25, 482–490. Doi: 10.1016/j.pt.2009.07.007.

[22] Mukherjee A, Padmanabhan PK. Sahani MH. Barrett MP. Madhubala R. Roles for mitochondria in pentamidine susceptibility and resistance in Leishmania donovani, Molecular and Biochemical Parasitology 2006;145, 1–10. Doi:10.1016/j.molbiopara.2005.08.016.

[23] Luque-Ortega JR, Rivas L. Miltefosine (hexadecylphosphocholine) inhibits cytochrome c oxidase in Leishmania donovani promastigotes. Antimicrobial Agents and Chemotherapy 2007;51, 1327–1332. Doi: 10.1128/AAC.01415-06.

[24] Azevedo Cardoso E, Rodrigues da Silva A, de Carvalho GC, Guerra Manssour Fraga A, de Castro Barbosa ML, Souza dos Santos AL, Castro HC, Lione V. Leishmaniasis: history, evolution of treatment and the need for new drugs. Current Biotechnology 2015;4279–4288. Doi:10.2174/2211550104666150303231736.

Chapter 4

Strategies Towards the Synthesis of Staurosporine Indolocarbazole Alkaloid and Its Analogues

B. Purna Chandra Rao, Osvaldo N. Oliveira Jr. and Ravi Varala

Additional information is available at the end of the chapter

http://dx.doi.org/10.5772/63832

Abstract

In this Chapter we revisit the main strategies used for years in synthesizing staurosporine indolocarbazole alkaloid and its analogues, which are promising compounds for treating cancer. In addition to describing the details of the synthesis strategies, including the key challenges that had to be faced, we offer a historical perspective of the development in the field.

Keywords: Indolocarbazole, alkaloids, cancer, synthesis, sugar moiety, glycosylation

1. Introduction

1.1. Aims and significance

Cancer is one of the most serious threats against human health [1], which has motivated extensive research into a plethora of chemotherapeutic agents [2-3]. The need for new anti-cancer drugs arises not only from the limitations of current drugs, but also from the development of drug resistance [4-6]. Several strategies exist for designing such novel drugs, for which the essential criterion is the selection of a suitable starting point from the vast chemical space [7]. Natural products, in this context, are privileged structures [8] and biologically prevalidated leads, for they contain molecules that probably evolved to exert highly specialized functions. About 74% of anticancer compounds originate from natural products or from natural product-derived products [9]. The variety of structures in products is key for new therapeutics [10].

The indolocarbazole family of natural products (hereafter referred to as ICZ's) was discovered in 1977 in actinomycetes, bacteria commonly found in soil, and is now investigated by medicinal chemists especially due to its antitumor and neuroprotective properties [11-13].

Figure 1 illustrates that ICZs are a structurally diverse family of natural products. The four types of aglycons include: A) the parent indolo[2,3-a]carbazole nucleus, such as that found in tjipanazole F2 (**7**); B) an imide, as in rebeccamycin (**8**) and arcyriaflavin D (**3**); C) hydroxy lactams, as in the UNC compounds (**2**); and D) simple lactams, found in **1**, **4**, **5** and **6**. In all of these aglycon types, substitution with halides, ethers, phenols, has been done at various positions on the aromatic heterocycle. The compounds possessing the pyrroloindolocarbazole system with one *N*-glycosidic bond, such as rebeccamycin (**8**), act by inhibiting *DNA topoisomerase* (target for cancer chemotherapy), whereas those with two *N*-glycosidic bonds, e.g., staurosporine (**1**), are mainly *protein kinase C (PKC) inhibitors* [14].

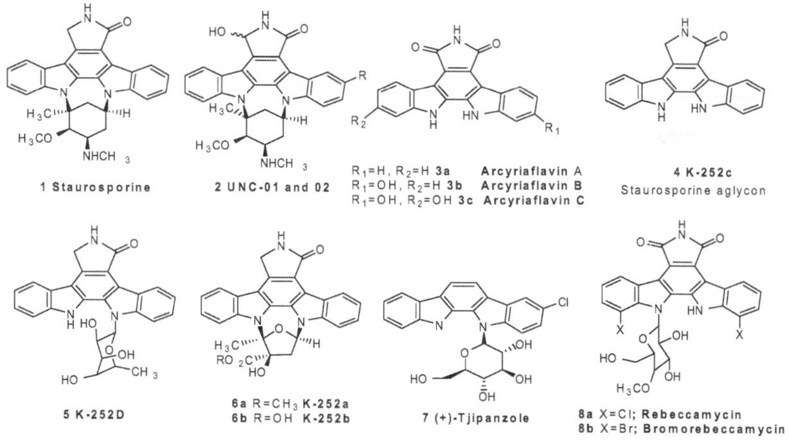

Figure 1. Well-known indolocarbazole alkaloids

We can also further divide ICZs based on the pattern of attachment of aglycon to the sugar moiety into four sub-patterns, viz.: A) ICZs having no sugar moiety, such as **3**, **4**; B) ICZs possessing one indole *N*-glycosidic linkages, such as **5**, **7** and **8**; C) ICZs with pyranose fused ring with two indole *N*-glycosidic linkages (e.g., **1**, **2**); and, D) ICZs with furanose fused ring with two indole *N*-glycosidic linkages (e.g., **6**). The synthetically most challenging subgroups of indolocarbazoles are the cyclofuranosylated [e.g., K252a (**6a**)] and cyclopyranosylated [e.g., staurosporine (**1**)] congeners.

Knolker and Reddy reviewed the synthesis and biological activity of carbazole alkaloids, depicted in Figure 2, where different synthetic strategies for indolocarbazole alkaloids were discussed [15].

1.2. Motivation for the chapter

Potent drugs against cancer normally have to fulfill a number of requirements in terms of its toxicity to tumor cells and solubility for efficient delivery. This requires a full-fledged charac-

http://dx.doi.org/10.5772/63832

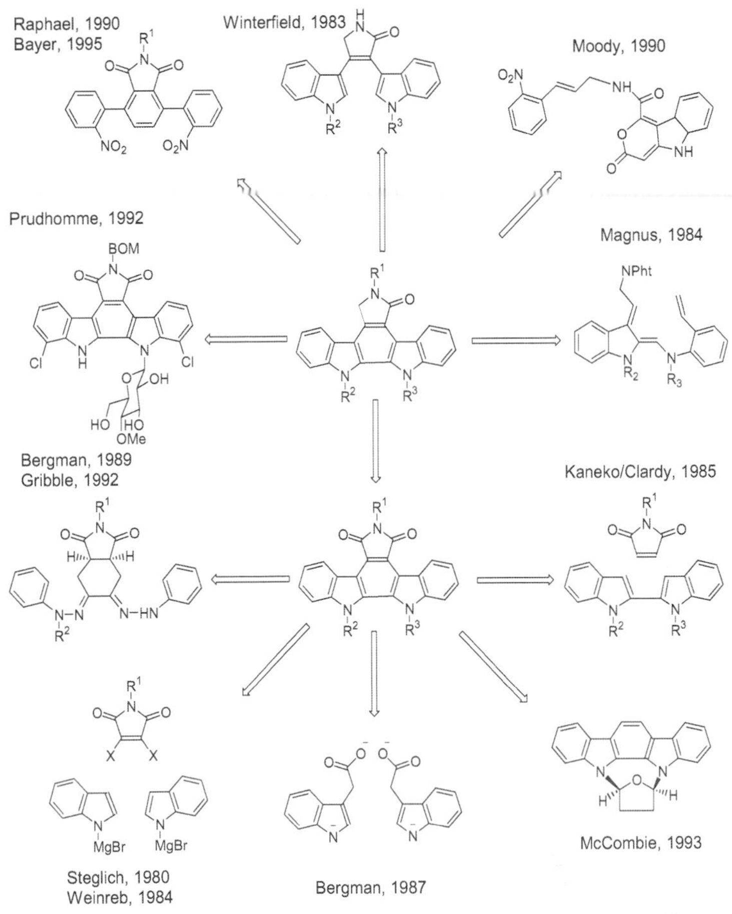

Figure 2. Approaches to indolocarbazole alkaloids

terization of drug candidates, including possible synthetic strategies. In this Chapter we concentrate on indolocarbazoles such as staurosporine, the most potent PKC inhibitors isolated to date, which probably act by occupying the ATP binding site and preventing protein phosphorylation. There is hence the need of synthetic routes to prepare indolocarbazole derivatives that are selective toward specific malfunctioning kinases associated with a disease. Furthermore, clinically useful compounds should have enhanced solubility in water, as compared to the poorly soluble ICZs. Since most indolocarbazoles with potent biological activities have substituents on the benzene portion of the core, enhanced solubility has been attempted with at least three approaches. The first is to introduce a hydrophilic group on the imide nitrogen, e.g. the *N*-bis(hydroxymethyl)methylamino group. The second possibility consists in elongating the carbohydrate side chain. The third approach is to replace the uncharged sugar residue of ICZ with a positively charged amino-carbohydrate. Many of the recent synthetic approaches toward indolo[2,3-]carbazole glycosides separately address the syntheses of the sugar and heterocyclic portions, leaving glycosylation as the consummate step. *One of the major difficulties associated with the synthesis of biologically-active ICZ alkaloids, such as Staurosporine, is the regiocontrol required for the glycosylation step.* Left undiscriminated to the last, the attachment of a chiral sugar moiety to a specific indolic nitrogen indolocarbazole moiety (R_1 =R_2 =H) occurs nonselectively, thus producing regioisomers.

Well-known examples of pharmaceutically important glycosylated natural products include macrolide antibiotics, aromatic polyketides, glycopeptides, indolocarbazoles, aminoglycosides, and cardiac glycosides. The sugar moieties are often essential for the biological activity in such natural products. Thus, altering the structures and/or substitution patterns of sugar appendages on aglycone moieties, a process known as *glycodiversification*, could potentially generate glyco-conjugates with enhanced biological activity. Therefore, glycodiversification may ultimately lead to new antibiotics against drug-resistant infectious bacteria, improved cytotoxic agents for treating cancer, or potent chemicals for combating other ailments.

1.3. Definition of the problem

The indolocarbazole acceptor is generally a weaker nucleophile than the bis(indoly1)-maleimide or indole acceptor, which limits application of established glycosylation methodologies to the indolocarbazole aglycones.

1.4. History of staurosporine

1.4.1. Isolation

Omura et al reported in 1977 a new alkaloid, isolated from *Streptomyces staurosporeus* during a search for new alkaloids in actinomycetes, found to possess potent hypotensive properties in addition to broad spectrum antifungal activity [16]. It was originally named as **AM-2282** (**1**), whose structure solved by single crystal X-ray analysis contained an indolocarbazole subunit with the two indole nitrogens bridged by glycosyl linkages (see Figure 3) [17-18]. AM-2282 was then renamed **staurosporine** (**1**), and became the first of over 50 compounds to be characterized in this family of alkaloids possessing the indolo[2,3-a]carbazole subunit [19-20].

Structure **1a**, the enantiomer of the natural product, was originally assigned to staurosporine, and not until recently has the assignment of the absolute configuration of staurosporine been revised to that shown in structure **1** (Figure 4) [21].

This isolation of staurosporine sparked research into related natural and synthetic compounds, particularly for treating cancer with nanomolar inhibition of protein kinases (PKC) [22]. Many staurosporine analogues are in phase III clinical trials to treat cancer and about ten such PKC inhibitors have been approved for use in clinical level.

http://dx.doi.org/10.5772/63832

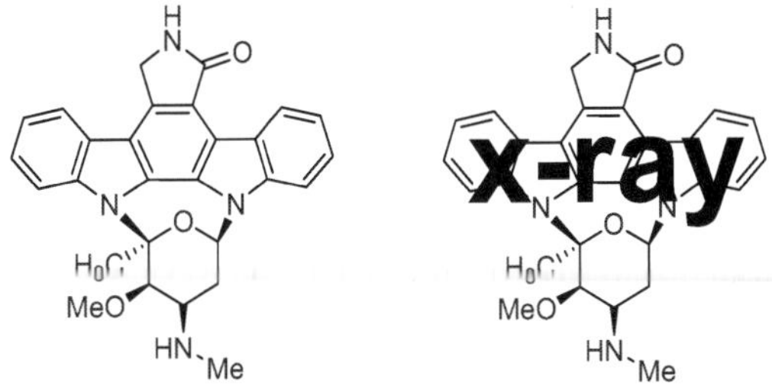

(+)-Staurosporine

Figure 3. Single crystal X-ray analysis of staurosporine

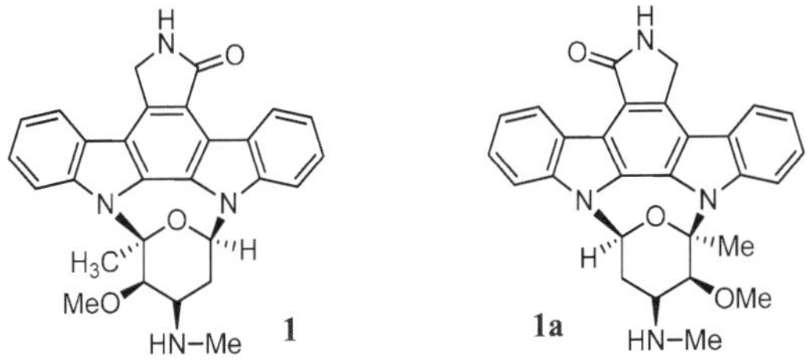

1 1a

Figure 4. Structures of staurosporine (**1**) and *ent*-staurosporine (**1a**).

1.4.2. The importance of protein kinase c inhibitors

Protein kinase C (PKC) is a family comprised of at least eight serine/threonine specific kinases that are approximately 77 kD in size. The importance of PKC in regulating signal transduction pathways and ultimately cellular response has been well-established [59]. Activation of PKC occurs through a series of events that begins with specific binding of an extracellular agonist to a cell surface receptor. This binding results in activation of phospholipase C which then cleaves inositol triphosphate (IP3) from phosphatidylinositol-4-5- biphosphate (PIP2) and leaves behind a molecule of 1,2-diacylglycerol (DAG) in the membrane. Phosphorylation ultimately results in cellular responses by modifying the function of rate-limiting enzymes and regulatory proteins implicated in metabolic pathways.

As already mentioned, indolocarbazoles such as K252a and staurosporine are the most powerful PKC inhibitors isolated to date. This mode of PKC binding, illustrated in Figure 5, unfortunately results in a relatively non-selective inhibition of several kinases. The preparation of indolocarbazole derivatives possessing selectivity toward specific malfunctioning kinases associated with a disease state would be a solution; thus, an efficient and general synthetic route to the indolocarbazoles is desirable.

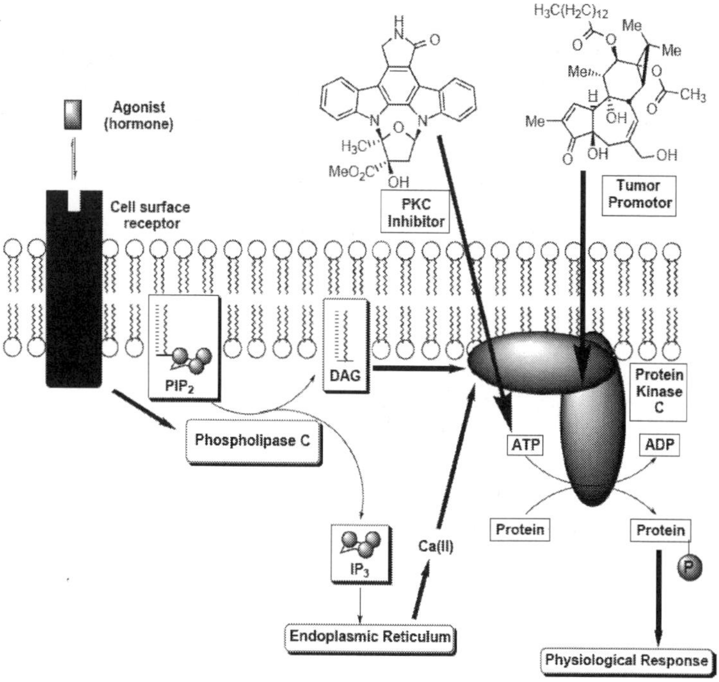

(Adapted from: B. M. Stolz, PhD Thesis, Yale University 1997)

Figure 5. Mechanism of PKC inhibitors

1.4.3. Pharmacology of staurosporine and its analogues

The recent literature on staurosporine analogues has provided valuable inputs into their biochemical pharmacology and generated discussion on the suitability of protein kinase C as potential target for anticancer drugs. The following conclusions are particularly pertinent with respect to pharmacological mechanisms [23]:

1. staurosporine analogues such as UCN-01 and CGP 41251 are inhibitors not only of PKC, but of a 'cocktail' of kinases;

2. the composition of this cocktail and expression of its constituent kinases in a given neoplasm determine the nature and extent of pharmacological efficacy; and

3. slight alterations in molecular structure dramatically alter individual components of this cocktail.

Indolocarbazoles are all biologically active and display such properties as antimicrobial, antifungal, and antitumor activity, in addition to acting as hypotensive or platelet aggregation agents [24-27]. Three representative examples of this class are staurosporine (**1**), rebeccamycin (**8**), and K-252a (**6**) (see Figure 1). Rebeccamycin (**8**) causes topoisomerase I-mediated DNA cleavage and is presently in late-stage clinical trials as an anticancer agent. Additionally, staurosporine (**1**) and K-252a (**6a**) are potential antitumor agents acting as potent inhibitors of protein kinase C (PKC). Staurosporine has also been reported to possess immunosuppressive activity and to reverse multidrug resistance [28-30]. It is because of its nanomolar inhibition of PKC, however, that staurosporine has attained its current acclaim.

2. Synthesis of staurosporine and its analogues

2.1. Introduction

Staurosporine can be divided into two distinct parts: the "northern" indolocarbazole aglycon and the "southern" carbohydrate portion of the molecule, as shown in Figure 6. One can envision that by so dissecting the molecule, a convergent synthetic approach would be possible in which a lactam-protected derivative of aglycon could be coupled with a bis-glycal derivative (no commitment is made as to the functional nature of R_1 or R_2).

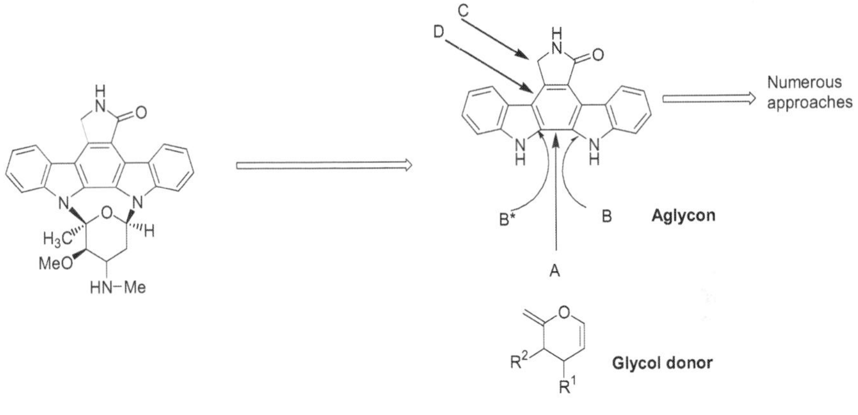

Figure 6. Retrosynthetic analysis of staurosporine (**1**).

From Figure 7 one may infer that aglycon **2** is itself a natural product, commonly referred to as staurosporinone or K-252c. Because it constitutes a major unit of many indolocarbazole natural products, several approaches to its synthesis have been developed [31-32]. Classified by the last covalent bond(s) formed, these approaches include cycloaromatization (A), double nitrene C-H insertion (B, B'), nitrene C-H insertion (B'), maleimide reduction (C), and diazo-lactam initiated cycloaromatization (D).

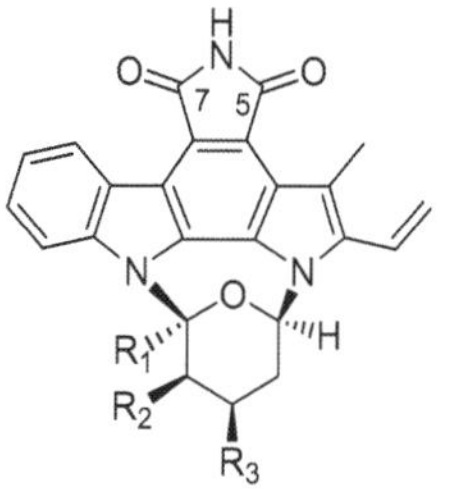

1a R_1, R_2=H: C5 and C7 are "enantiotopic"
1b R_1, R^1H: C5 and C7 are regiodifferentiated

Figure 7. Effect of substitution on differentiation of C5 and C7.

2.2. Biosynthetic pathway of staurosporine

2.2.1. Biogenesis of the indolocarbazole nucleus

Cordell and Pearce independently reported the first indolocarbazole biosynthesis in 1988 [33-35], both identifying aglycon units of ICZs (**1** and **8** (Figure 1)), to be derived from two intact tryptophan units. Tryptophan (**10**) was in fact utilized in the aglycon biosynthesis, produced by Streptomyces staurosporeus from D-glucose (**9**), probably via the shikimic acid pathway (Scheme 1) [36].

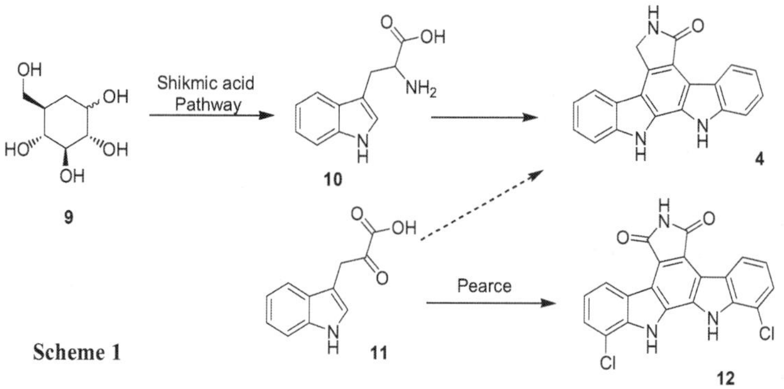

Scheme 1

2.2.2. Biosynthesis of indolocarbazole carbohydrates

The carbohydrate precursor to staurosporine has been shown to be D-glucose and the *N*- and *O*-methyl groups are derived from L-methionine as shown in Scheme 2. Hoehn reported the isolation of **15b** by cofermentation and bioconversion studies and found that *O*-methylation is the last step, ie., direct precursor to staurosporine biosynthesis [37].

Scheme 2

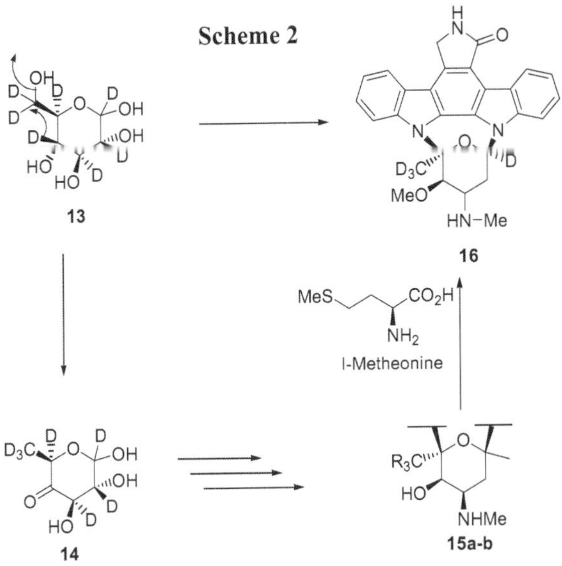

2.2.3. *About this pathway*

The first enzyme identified in staurosporine biosynthesis was the one catalyzing the very last step (3'-*O*-demethyl-staurosporine methyltransferase). A *Streptomyces longisporoflavus* mutant defective in this enzyme was reported in 1995 [37], while the enzyme was identified in 1998 [38]. The complete staurosporine biosynthetic gene cluster was cloned from *Streptomyces sp.* L-amino acid oxidase staO initiates synthesis by converting L-tryptophan to the imine form of indole-3-pyruvate (2-imino-3-(indol-3-yl)propanoate). StaD (staD) then catalyzes coupling of two IPA imines to yield chromopyrrolate. Formation of the indolocarbazole core of staurosporine is catalyzed by two enzymes: staP converts chromopyrrolate into three indolocarbazole compounds, K-252c, 7-hydroxy-K252c and arcyriaflavin A, by intramolecular C-C bond formation and oxidative decarboxylation, while StaC is required to ensure that the main product is K-252c.

The next step is glycosylation, which is catalyzed by two enzymes. K252c *N*-glycosyltransferase (staG) catalyzes *N*-glycosidic bond formation between N-13 and C-6' of the nucleotide sugar dTDP-L-ristosamine. Cytochrome P450 StaN (staN) then catalyzes an additional C-N bond formation between N-12 and C-5'. These two enzymes convert K-252c to 3'-*O*-demethyl-4'-*N*-demethyl-staurosporine *via* the intermediates holyrine A and holyrine B. The final steps in the pathway are two methylation reactions. staMA catalyzes *N*-methylation of 3'-*O*-demethyl-4'-*N*-demethyl-staurosporine and staMB catalyzes *O*-methylation, which results in staurosporine (**Figure 8)** [39].

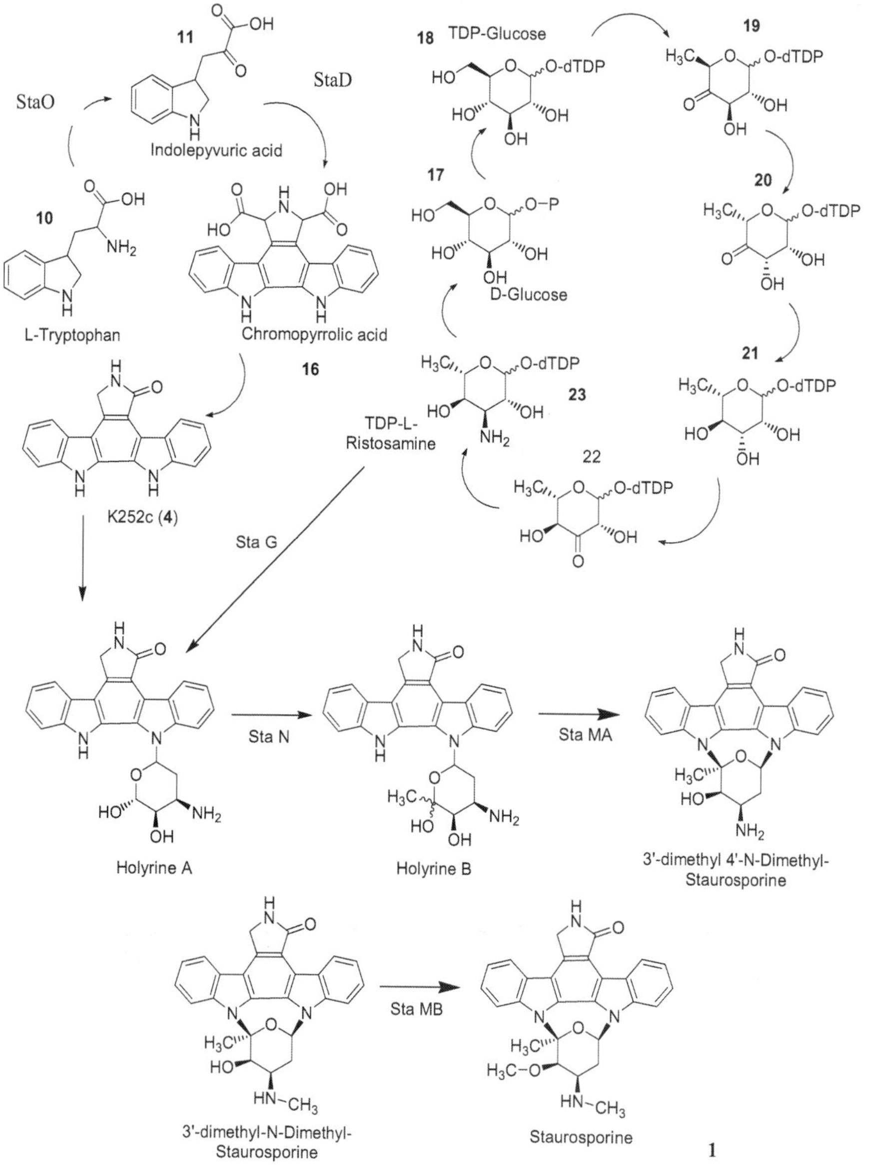

Figure 8. Biosynthetic pathways of Staurosporine in Stryptomyces sp. TPA0274. The genes associated in synthetic steps are shown (dTDP-deoxy-thymidine-5-diphosphate).

http://dx.doi.org/10.5772/63832

2.3. First total synthesis of staurosporine and ent-staurosporine (Danishefsky et al., 1995)

It was not until 1995 that the first total synthesis of staurosporine (**1**) was reported by Danishefsky et al. [40]. A central challenge in total synthesis by previous groups was that of constructing the two glycosidic bonds to weakly nucleophilic indolic nitrogens [41-44]. Danishefsky observed oxazolidinone glycal **27b** to function as the glycosyl donor and bis(indolyl)maleimide **26** to function as the aglycon acceptor (Scheme 3). Aglycon **26** was synthesized from benzyloxymethyl (BOM) dibromomaleimide **24** in the modular fashion shown.

Scheme 3

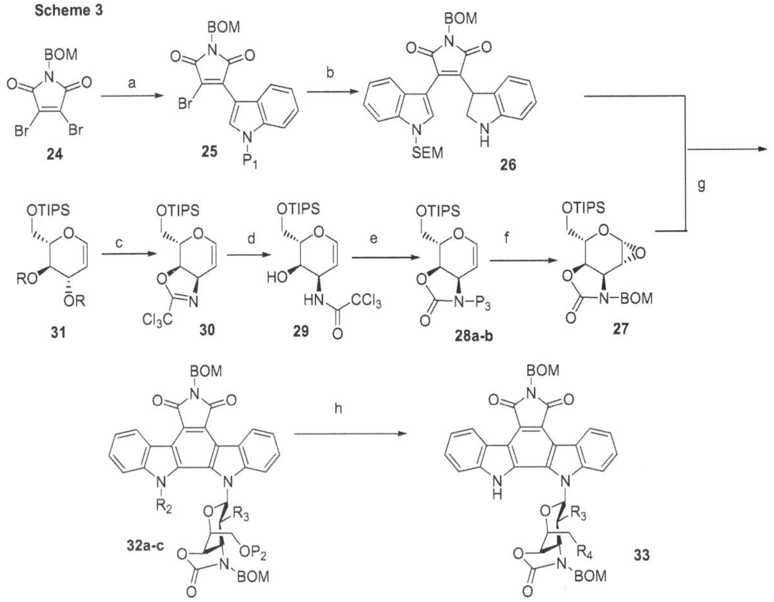

(a) (i) Indole Grignard, PhH 0 °C to rt, overnight, 82%(P_1 = H). (ii) NaH, THF, room temp., then SMECl, 91% (P_1= SEM). (b) Indole Grignard, PhH 0°C to rt, overnight, 75%. (c) NaH, CH_2Cl_2, 0°C then Cl_3CCN,0°C to rt,(R= $CNHCCl_3$ then BF3.OEt2 -78°C, 78%. (d) cat. TsOH,H_2O, pyr, 80°C, 80%. (e) (i) NaH, CH2Cl2, 0°C to rt, 92% of 27 (P_2 = TIPS, P_3=H)(ii) NaH, DMF, then BOMCl, 40 °C, 65%((P_2 = TIPS, P_3=BOM) and 22% of 27 (iii) TBAF, THF, 0 °C 95%, (P_2 = H, P_3=BOM) (iv) NaH, DMF, 0°C to rt then PMBCl, 0°C to rt 92% of 28 (P_2 = PMB, P_3=BOM) (f) Dimethyldioxirane, CH_2Cl_2, 0 °C 100%, of α–epoxide β-epoxide. (g) (i) 21, NaH THF, rt then 11 and 12, rt to reflux 47%of 30 (P_2 = PMB, R_2=SEM R_3 =OH) (ii) Thiophosgene, DMAP, Pyr, CH_2Cl_2 reflux then C_6F_5OH reflux, 79% (P_2 = PMB, R_2=SEM R_3 =$OCSOC_6H_5$ (iii) n-Bu_3SnH, AIBN, PhH reflux, 74% 31 (P_2 = PMB, R_2=SEM R_3 =H) (iv) DDQ CH_2Cl_2 H_2O 0 °C to rt 97% (P_2 = H, R_2=SEM R_3 =H) (v) TBAF, THF reflux, 91% of 32 (P_2 = H, R_2=H R_3 =H) (h) (i) hυ, cat. I2, air, PhH, rt 73% of 16 (R_4 =OH). (ii) I_2, PPh_3, imidazole, CH_2Cl_2, 0 °C to rt 84% (R^4 =I)

Triisopropylsilyl-L-glucal **31** (TIPS-L-glucal) was converted to its bis- (trichloroacetimidate) and then to oxazoline **30** by an apparent vinylogous Schmidt glycosylation. The oxazolidinone, fashioned from derivative **29**, was protected as its BOM derivative **27**. The TIPS protecting group was cleaved, and a *p*-methoxybenzyl ether (PMB) was installed. Accordingly, glycal **28b** was oxidized with 2,2-dimethyldioxirane. The mixture of epoxides (**27**) was treated with the sodium salt of **26** to furnish indole glycoside **32** with 47% yield.

Compound **32a** was subjected to Barton deoxygenation to remove C_2' hydroxyl, affording **32b**. Seco-system **32c** was obtained by further deprotection of C_6' PMB and the indolic SEM groups. Photolytic oxidative cyclization resulted in compound **33** (Scheme 3). The exo-glycal, which was essential for intramolecular glycosylation, was performed using iodination strategy of **33** followed by elimination. Treatment of **34** with potassium *tert*-butoxide and iodine eventually resulted **35.** Thereafter, reacting with tri-*n*-butyltin hydride and deprotecting the BOM groups, compound **37** was available. For compound **38**, a BOC group was introduced particularly on the oxazolidinone ring to facilitate disconnection of oxazolidinone. The BOC group would protect against dimethylation of the amine during the opening reaction. To safeguard the imide ring during sequential modifications, which would generate *N*-methyl and *O*-methyl functions, compound **38** was converted into **39**. Treatment of **39** with cesium carbonate in methanol led to **40**. Next, the *O*-methyl and single *N*-methyl groups were incorporated to yield **41**, which on further deprotection afforded 7-oxostaurosporine (**42**) (Scheme 4). 7-Oxo compound **42** was transformed into staurosporine.

A methodology was developed to convert the 7-oxo compound **42** to staurosporine itself. It started with a reduction with sodium borohydride (In Scheme 4, 40-42). It was not that easy to deoxygenate the carbanolamide linkage but this portion was smoothly accomplished by using benzeneselenol. By performing two steps on **42**, Danishefsky et al. obtained a 1:l mixture of isostaurosporine (**43**) and staurosporine (**1**) [40]. After separation, homogeneous fully synthetic staurosporine (**1**) was isolated. The total synthesis of staurosporine (**1**) was thus completed.

2.4. Staurosporine and ent-staurosporine: The first total syntheses, prospects for a regioselective approach, and activity profiles (Danishefsky et al., 1996)

The total syntheses of staurosporine and *ent*-staurosporine was achieved again by Danishefsky et al, by constructing both the glycosidic bonds from glycal precursors [45]. The first glycosidic bond was originated from direct epoxidation of endo-glycal to give 1,2-anhydro sugar, which was later made to react with indole anion through intermolecular coupling. They used the strategy of intramolecular iodo glycosylation for the second bond using an exo-glycal [45].

The authors dealt with the problem of indole glycosylation, functional group management in the pyranose ring, and regiochemical harmonization in the course of the first total synthesis of staurosporine (**1**) detailed herein. *It is an electrophilically induced cyclization of the second indolic nitrogen onto a novel exo-glycal to establish the staurosporine core skeleton.*

Monosaccaharide synthesis

Danishefsky et al. assumed the upcoming C_3' methoxy and C_4' methylamino vestiges would be existing in an oxazolidinone ring. Protecting the nitrogen with a benzyloxymethyl group, C_1'-*p*-methoxybenzyl ether would protect a primary alcohol that could be utilized in designing the exo-glycal essential for intramolecular indole glycosylation (Scheme 5).

Scheme 4

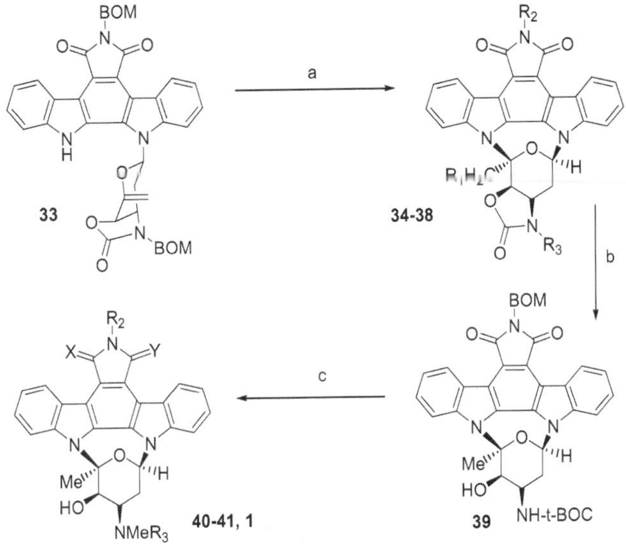

16 (R_4 =I), THF, DBU, roomtemperature, 89% of 7. (a) (i) t-BoOK, I_2, THF, MeOH, room tempeature, 65% of 18 (R_1= I, R_2, R_3 = BOM) and 15% of recovered 17. (ii) n-Bu3 SnH, SnH, AIBN, PhH reflux 99% of 19 (R_1= H, R_2, R_3 = BOM) (iii) H2, Pd, (OH)2, EtOAc, MeOH, room temp. then NaOMe, MeOH, 90% of 20(R_1= R_2=R_3 = H) (iv) $(BOC)_2O$, THF, cat. DMAP, room temp. 81% of 21 (R_1= = H, R_2, R_3 = BOC). (v) NaH, DMF, MeOH, room temperature, then BOMCl, 82% of 22 (R_1=HI, R_2= BOM R_3 = BOC). (b) Cs_2CO_3, MeOH, room temp. 93%. (c) (i) NaH, $(CH_3)_2 SO_4$, THF, DMF, room temp. 86% of 24. (X, Y=O R_2=BOM R_3 = BOC). (ii) H_2 $Pd(OH)_2$, EtOAc, MeOH, room temp. then NaOMe in MeOH 84% (X, Y=O R_2=H R_3 = BOC) (iii)TFA, CH2Cl2, room temp. 97% of 25 (X, Y=O R_2,R_3 = H) (iv) NaBH4, EtOH, room temp. work up (X, Y= O, OH R_2, R_3 =H), then PhSeHcat. TsOH, CH_2Cl_2 room temp. 39% of (X=H2 Y=O R_2, R_3 =H) 39% of 26 (X=O, Y=H2 R_2, R_3 = H), and 5% of 25.

Consistent with the discussion above, they formulated the donor to be a glucal of the type **50a**. This sequence shown in Scheme 5 provided oxazolidinone glycal **50a** which proved to be an effective glycosyl donor subjecting to proper activation. It was noted that oxazolidinone would provide stereochemical guidance in activating the endo-glycal *en route* to the first indole glycosylation.

Glycosylation and Elaboration.

Oxazolidinone glycal **50a** and its derived epoxide proved to be effective as functional versions of target glycals. Danishefsky et al. next focused on the first glycosidic bond (Scheme 6), for which bis-indolyl maleimides were effective glycosyl acceptors for 1,2-anhydrosugar donors. Thus, the sodium anion of bis-indolyl maleimide **26** was synthesized and treated with a solution of 1,2-anhydrosugars prepared from epoxidation of glycal **50a** using 3,3-dimethyl-dioxirane. A mixture of expected indole glycoside **52** (47% yield) and indole glycoside **53** (10% yield) was obtained upon heating the reaction.

Scheme 5

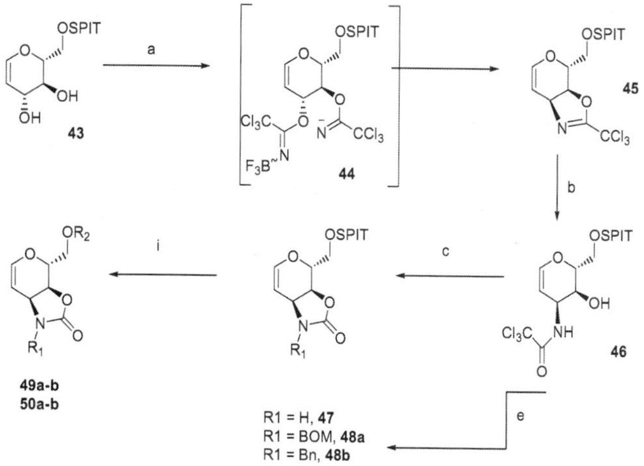

a) NaH, CH_2Cl_2 0 °C then Cl_3CCN, 0 °C to rt then BF_3. OEt_2 -78°C to 78% (b) TsOH H_2O, pyr, 80°C, 80% (c) NaH, CH_2Cl_2 0 °C to rt, 92% (d) NaH, TBAI, DMF then BMOCl, 40 °C 65% and 22% recovered 48. (e) NaH, TBAI, DMF, then BnBr, 0 °C to rt, 94% (f) TBAF, THF, 0 °C , 95% (g) NaH, DMF, 0 °C to rt then PMBCl, 0 °C to rt, 92%

Scheme 6

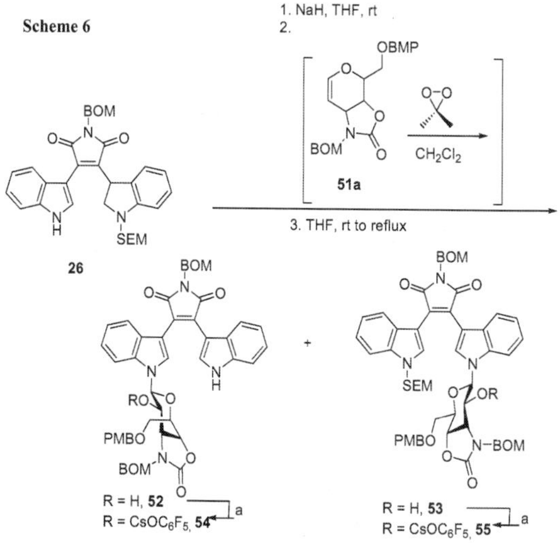

(a) Thiophosgene, DMAP, Pyr, CH_2Cl_2 reflux then C_6F_5OH, reflux, 79%

http://dx.doi.org/10.5772/63832

Alteration of the functional group was essential to construct the second glycosidic bond. It was performed by deoxygenating the newly created alcohol at C5′, deprotecting the indole moiety, establishing 2,2′ indolic bond, and finally formation of exo-glycal (Scheme 7).

The Key Cyclization

Early screening of the reaction of indolocarbazole glycoside **61** with an array of electrophiles failed to establish conditions to perform cyclization and lead to the fully functionalized core of staurosporine (**1**). Indolocarbazole glycoside **61** should have its activated exo-glycal and thereby undergo a conformational change so that cyclization would be made possible. The sterically demanding aglycon must be in an axial conformation rather than the preferred equatorial conformation. Cyclization to **62** thus resulted as the nucleophillic nitrogen attacks the activated exo-glycal.

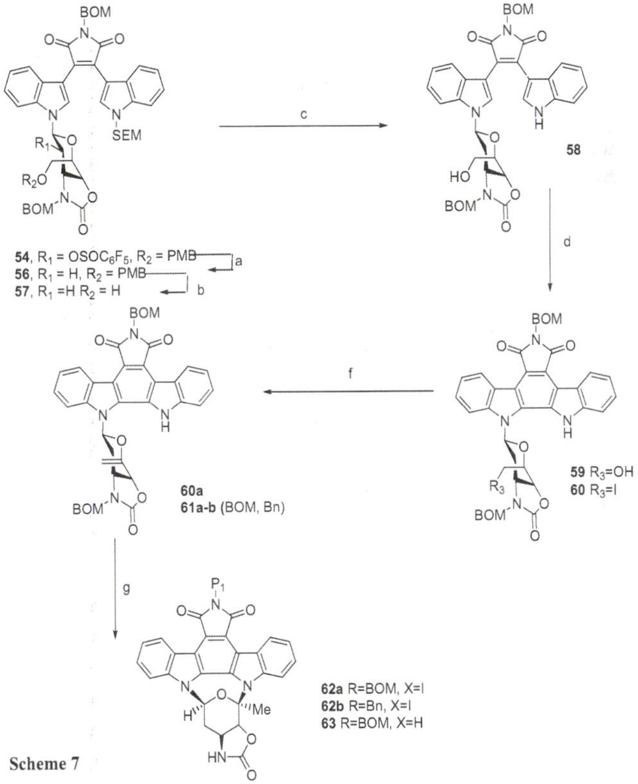

Scheme 7

(a) n-Bu_3SnH. AIBN ,PhH, reflux, 74%. (b) DDQ, CH_2Cl_2, H_2O, 0 °C to rt, 97% (c) TBAF, THF, reflux, 91%. (d) hυ, cat. I_2, air, PhH, rt, 73% (e) I_2, PPh_3, imidazole, CH_2Cl_2, 0 °C to rt, 84% (f) DBU, THF, 0 °C, 89%; (h) KOtBu, I_2, THF:MeOH, rt.

Completion of the Synthesis

To complete the total synthesis of *ent*-staurosporine (**2**), a two-step deprotection strategy (hydrogenation followed by aminal hydrolysis) delivered **64** from **63** in high yield (Scheme 8). Danishefsky et al. preferred to clarify the monosaccharide domain prior to reducing the maleimide function [45]. The most efficient method involved reduction of the imide group with sodium borohydride to provide a 1:1:1:1 mixture of hydroxy lactams. Further reduction to *ent*-staurosporine (**1a**) and *ent*-isostaurosporine (**71**) was then successfully finished using phenylselenol and *p*-TSA. Compounds **1a** and **71** were each isolated in a homogeneous state from the 1:1 mixture generated from this two-step sequence.

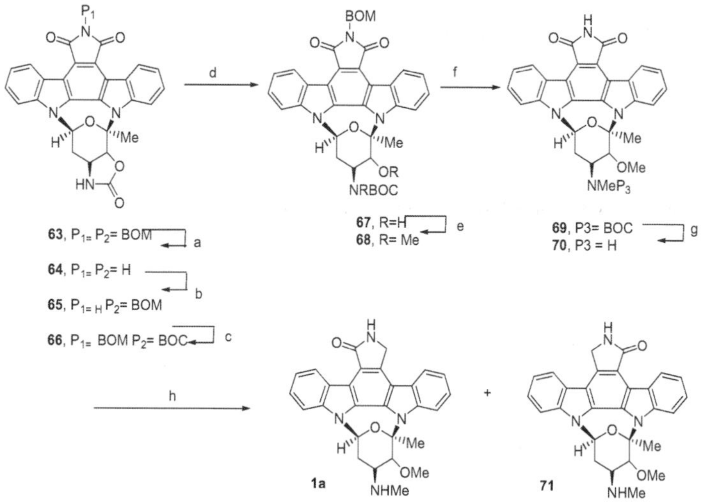

Scheme 8

(a) H_2, $Pd(OH)_2$, EtOAc, MeOH, rt, then NaOMe in MeOH, 92% (b) BOC_2O, THF, cat. DMAP, rt, 81% (c) NaH, DMF, rt, then BOMCl, 82% (d) Cs_2CO_3, MeOH, rt, 93% (e) NaH, $CH_3)_2SO_4$, THF, DMF, rt, 86% (f) H_2, $Pd(OH)_2$, EtOAc, MeOH, rt, then NaOMe in MeOH, 84% (g) TFA, CH_2Cl_2, rt, 97%, (h) $NaBH_4$, EtOH, rt, workup, then PhSeHcat. TsOH, CH_2Cl_2, rt 39% of 2, 39% of 89 and 15%of 88

Upon successfully completing the chemistry in the *ent*-series, the strategy towards total synthesis of staurosporine (**1**) was evident (Scheme 9). Initially, tri-*O*-acetyl-L-glucal **72** was transformed into the corresponding oxazolidinone **73**. Compound **74** resulted from coupling to the aglycon, deoxygenation, photocyclization, and finally by exposing exo-glycal. Thereafter, performing the crucial cyclization step yielded **75**. Opening the oxazolidinone, methylation, deprotection, and reduction furnished staurosporine (**1**) and isostaurosporine (**1a**).

Danishefsky et al. evaluated *ent*-staurosporine (**1a**), *ent*-isostaurosporine (**71**), a related imide **64**, and their corresponding enantiomers for their *in vitro* antitumor activity, their capacity to inhibit PKC (Table 1), and their ability to inhibit topoisomerase I. The cytotoxicity of indolo-

http://dx.doi.org/10.5772/63832

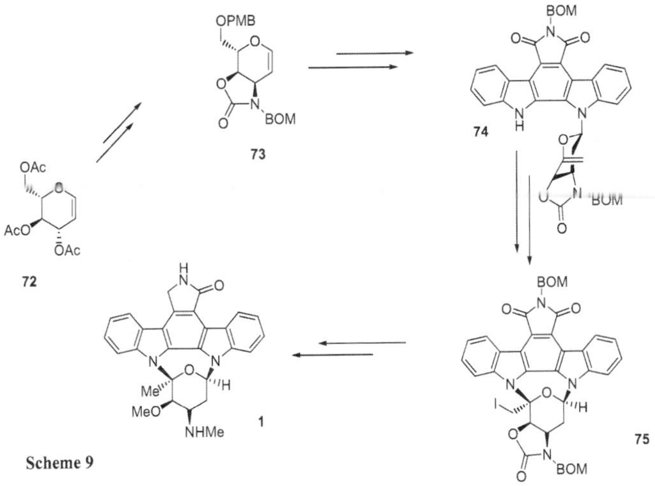

Scheme 9

carbazole alkaloids can also be affected by a different mechanism than inhibition of PKC, i.e. inhibition of topoisomerase I (Table 2).

Compound	PKC inhibition: IC_{50} (μM)		Cytotoxicity HL-60	Cytotoxicity 833K
Staurosporine (**1**)	α	0.02	0.014	0.0065
	$β_1$	< 0.005		
	$β_2$	< 0.005		
ent-Staurosporine (**2**)	α	0.29	3.84	0.312
	$β_1$	0.19		
	$β_2$	0.15		
isostaurosporine (**94**)	α	0.98	2.39	0.55
	$β_1$	0.08		
	$β_2$	0.05		
ent-isostaurosporine (89)	α	0.03	6.55	0.272
	$β_1$	0.02		
	$β_2$	0.01		
imide(95)(*ent*-82)	α	1.25	1.68	0.72
	$β_1$	0.25		
	$β_2$	0.10		
imide **82**	α	0.24	593	454
	$β_1$	0.02		
	$β_2$	0.03		

Cytotoxicities are given as IC_{50}'s in μM units

Table 1. PKC Inhibition and in Vitro Cytotoxicity

Compound	topo I inhibition	
	DNA cleavage	inhibition of supercoiled DNA relaxation
Staurosporine (**1**)	++	+
ent-Staurosporine (**2**)	++	+
isostaurosporine (**94**)	++	+
ent-isostaurosporine (89)	+++	+
imide(95)(*ent*-82)	++	++++
imide **82**	+	+
camptothecin	++++	++++

Relative potencies are compared with camptothecin (++++) at 100μM

Table 2. Topo I Inhibition

2.5. Wood and Stolz's synthesis of staurosporine

A total synthesis of the natural product (+)-staurosporine has been achieved [46] along with other ICZs. The synthetic strategy involved steroselective ring expansion of a furanosylated indolocarbazole [(+)-79] to a pyranosylated congener [(+)-80] that serves a common intermediate in the production of 1 and other desired ICZs.

2.5.1. Retrosynthetic analysis: The development of a ring expansion approach to the pyranosylated indolocarbazoles

Wood and Stolz began by considering approaches that involved ring expansion of a furanosylated intermediate. Noting the striking structural homology of **1** and other related ICZs, they envisioned a strategy that would allow access to these congeners via a common intermediate. Specifically, R-methoxy ketone **76** was viewed as ideal since the stereogenic centers common in place and flexibility for stereocontrolled functionalization at C(4′) and C(5′) is maintained. Thus, reductive amination would produce staurosporine (**1**).

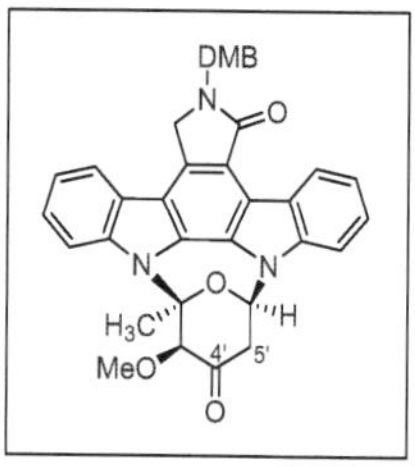

Figure 9. Key intermediate **76**

http://dx.doi.org/10.5772/63832

The inspiration for developing this approach derived from Wood's recognition that ketone **76** might be accessed from aldehyde **79** *via* a Tiffaneu-Demyanov-like ring expansion (Schemes 10 and 11). In designing this ring expansion approach, Wood et al. addressed the issues of regio- and stereochemical outcome and the known propensity of similar systems to undergo skeletal rearrangement (i.e., **77** to **78**, Scheme 10). From Scheme 3, it could be envisioned that the planned rearrangement occurs with migration of either bond *a* or bond *b* of aldehyde **79**, to produce regioisomeric hydroxy ketones **80** or **81**, respectively. Thus, Wood et al [46] assumed bond *a* would migrate to the *si* face of the aldehyde, producing a product (**80**) that possesses both the regio- and stereochemistry needed for further progressive steps towards staurosporine.

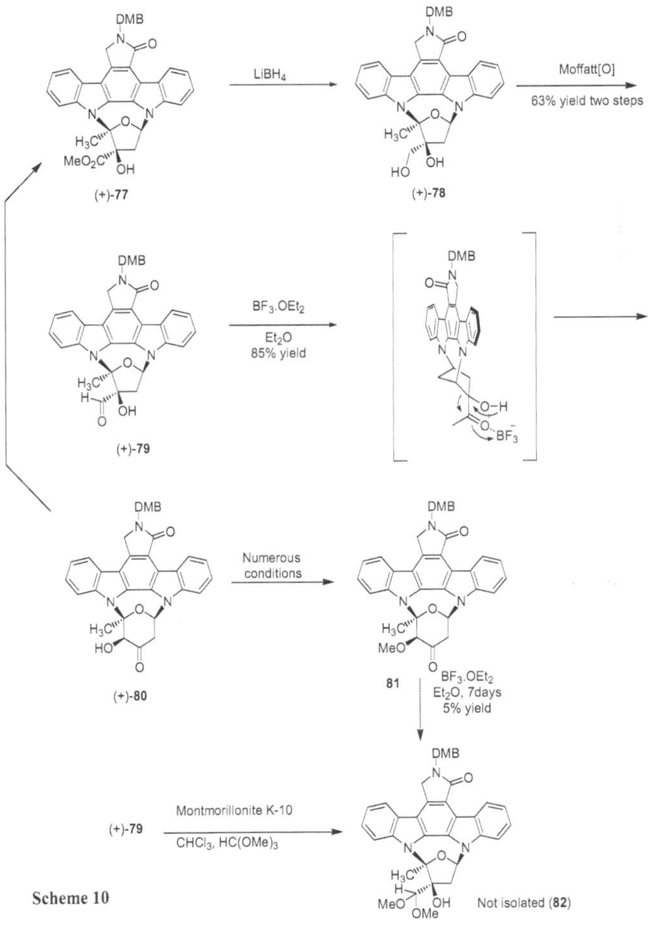

Scheme 10

2.5.2. Completion of staurosporine

Next, Wood and Stolz [46] treated (+)-**80** with $NH_2OH.HCl$ to produce corresponding oxime (-)-**83** in 95% yield. In contrast to ketone (+)-**80**, bis-methylation of (-)-**83** under phase transfer conditions (MeI, KOH, and n-Bu_4NBr in THF) occurred cleanly to afford (-)-**84** and set the stage for a stereoselective reduction (H_2/PtO_2) that furnished amine (+)-**85a**. Mono-methylation and deprotection then afforded (+)-staurosporine (**1**) in 67% yield (two steps, Scheme 11).

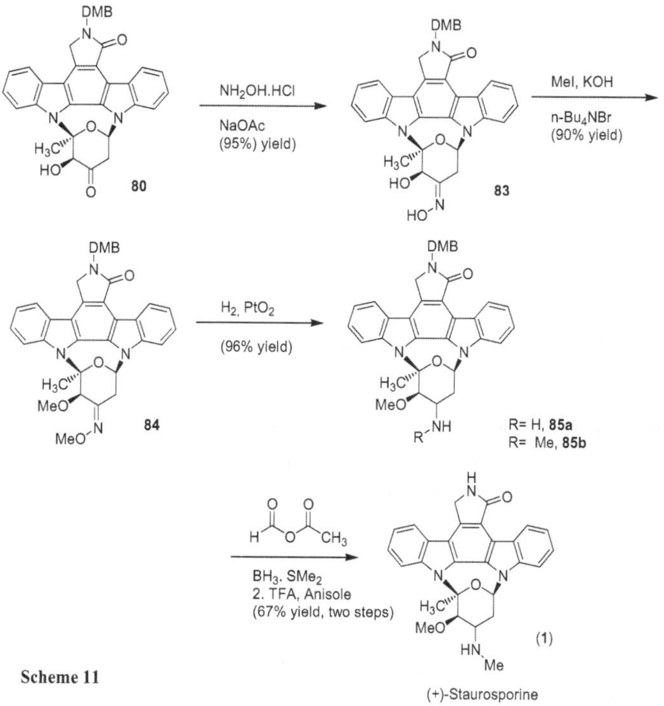

Scheme 11

3. The synthesis of carbohydrates for indolocarbazole synthesis

Only a few methodologies have been developed for synthesizing complex carbohydrate intermediates for use in the total synthesis of indolocarbazole alkaloids such as staurosporine (**1**). Some of those strategies are summarized in the succeeding sections:

3.1. Synthesis of staurosporine monosaccharide (Weinreb et al.)

Weinreb published the synthesis of aminohexose fragment of staurosporine *via* an *N*-sulfinyl Diels-Alder [4+2] cycloaddition [43,47]. From Scheme 12, cycloaddition of diene **86** and benzyl sulfinylcarbamate (**87**) resulted in a mixture of diastereomeric sulfoxides which after oxidation yielded the corresponding sultam (**88**) and then converted to acetal **99.** Subjecting to diastereoselectively epoxidition of olefin **89** using trifluoroperacetic acid afforded **90**. Hydrolytic-reductive opening of epoxide **90** followed by olefin cleavage resulted in keto-acetal **91**, a critical synthon for the staurosporine carbohydrate (**92**).

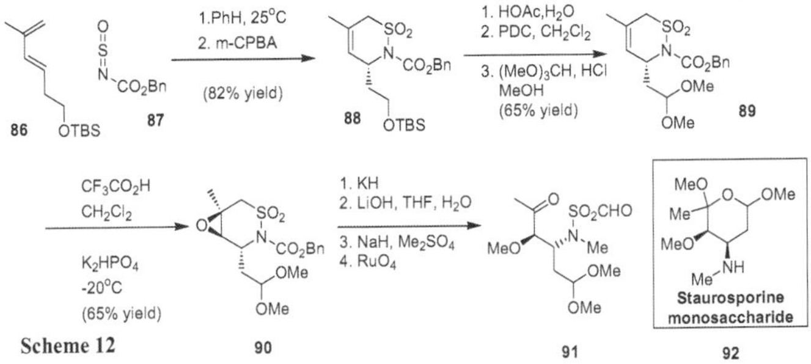

Scheme 12

3.2. Staurosporine glycal precursor (Danishefsky et al).

Danishefsky exploited glycal epoxide **93** as the glycosyl donor in his first total synthesis of staurosporine [41,48]. Glycal **94**, a derivative of L-glucal, was transformed into its corresponding oxazoline **95** by a modified Schmidt reaction. Conversion to oxazolidinone proceeded under standard conditions, and finally treatment with Murry's reagent provided the glycal epoxide (**96**, Scheme 13).

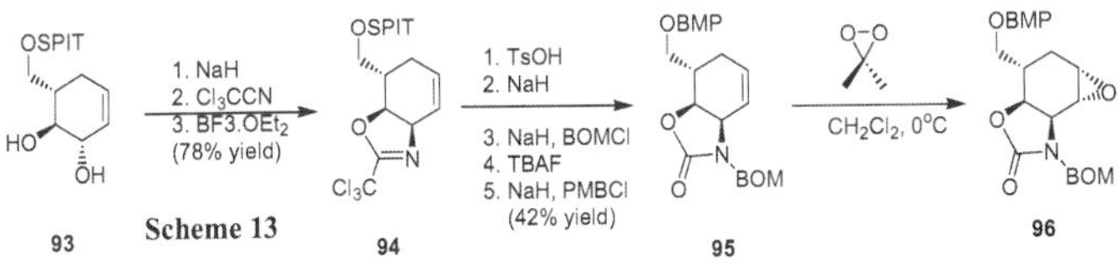

Scheme 13

3.3. Methods describing the combination of carbohydrate and indolocarbazole

3.3.1. *The Danishefsky synthesis of (+)- and (-)-staurosporine*

Danishefsky formulated a strategy to staurosporine [41], in which epoxidation of glycal (-)-**98** with maleimide **97** resulted in one of the indole *N*-glycosidic linkages to form **99**. Treatment of olefin **99** using Barton deoxygenation, iodine and *t*-BuOK followed by radical dehalogenation provided the pyranosylated indolocarbazole **101** with 64% yield. Deprotection and methylation followed as shown in Scheme 14 (i.e., **101**→**102**), after which reduction of imide **102** led to a 1:1 mixture of **1** and **1a**.

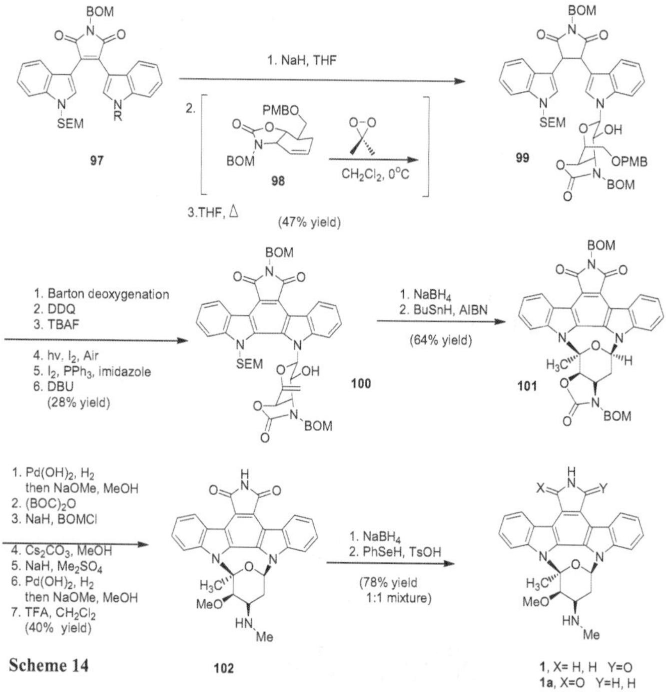

Scheme 14

3.3.2. *Syntheses, biochemical and biological evaluation of staurosporine analogues from the microbial metabolite rebeccamycin*

To synthesize staurosporine analogues from rebeccamycin, different structural variations were exploited by Prudhomme et al., including coupling of the sugar moiety to the second indole nitrogen, dechlorination and then reduction of imide to amide [49].

The synthesized compounds **105-109** in Scheme 15 were tested for their ability to bind to DNA and inhibit topoisomerase I and protein kinase C [49]. The cytotoxicity of dechlorinated imide analogue **108** correlates well with its DNA binding and anti-topoisomerase I activities.

3.3.3. *Synthetic studies on indolocarbazoles: Total synthesis of staurosporine aglycon*

Mohankrishnan et al synthesized staurosporine aglycon and its analogues with 28-36% overall yield, using 2-methylindole (**110**) as synthetic precursor [50]. The key steps for the synthesis of indolocarbazole alkaloids involved electrocyclization and nitrene insertion reactions as depicted in Schemes 16 and 17.

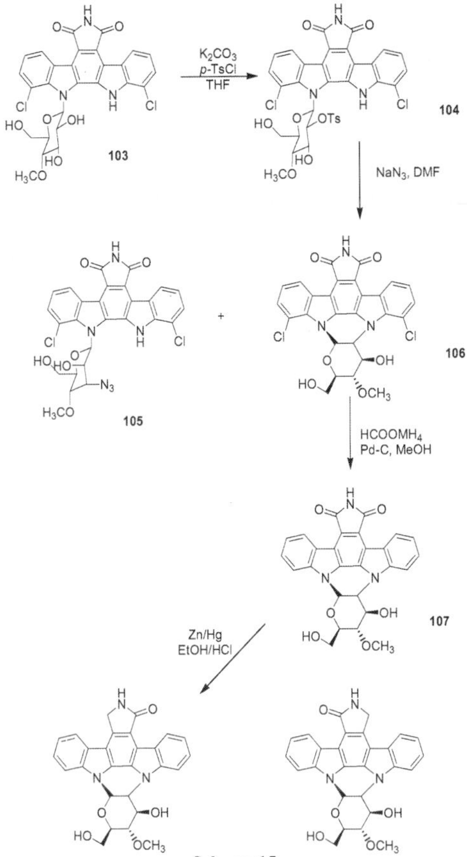

Scheme 15

Triphenylphospite-mediated nitrene insertion of 2-nitroarylcarbazole was performed at a moderate temperature using anhydrous $ZnBr_2$ as catalyst. In addition, an alternative synthetic protocol for preparing ICZs involving concurrent electrocyclization followed by nitrene insertion was adopted as in Scheme 17 by Mohankrishna et al. [50].

3.3.4. Synthesis of pyrrolidin-2-ones and staurosporine aglycon (K-252c) by intermolecular Michael reaction

3,4-Disubstituted pyrrolidin-2-ones, a group of compounds with interesting biological properties, are related to staurosporinone. The most important property is inhibition of protein kinase C (PKC), so that this antiproliferative agent can interfere with the cell cycle. The synthetic strategy permits preparation of said compounds using an intermolecular Michael addition, starting from nitroethene derivatives and substituted acetate Michael donors [51].

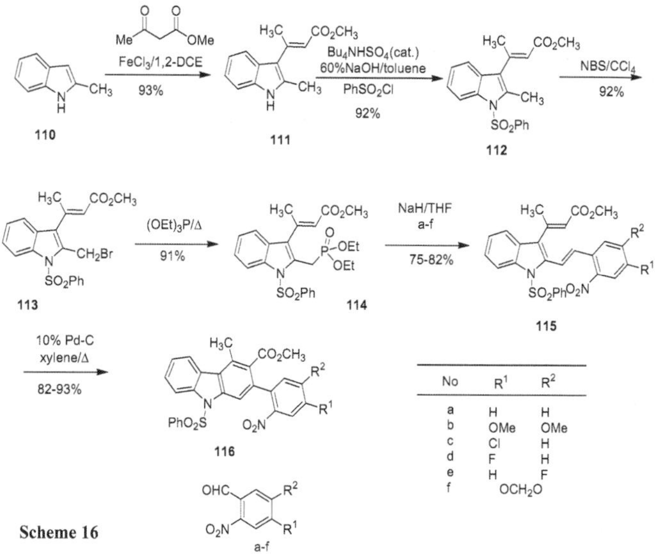

No	R^1	R^2
a	H	H
b	OMe	OMe
c	Cl	H
d	F	H
e	H	F
f	OCH_2O	

Scheme 16

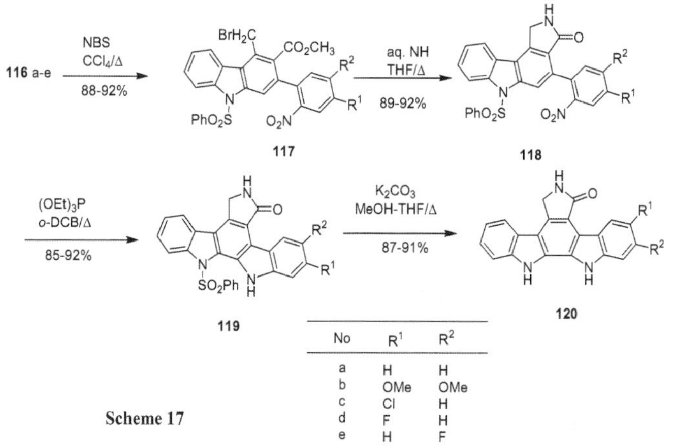

No	R^1	R^2
a	H	H
b	OMe	OMe
c	Cl	H
d	F	H
e	H	F

Scheme 17

Enantioselective syntheses can also be carried out using chiral auxiliaries in this strategy. Reduction of the nitro group using raney nickel and subsequent lactamization, the desired lactam precursor of staurosporine, which is essential for the biological activity, is obtained according to Scheme 18. The easiest and shortest (in contrast to the published routes of staurosporinone) synthetic strategy of staurosporinone within three steps with good to moderate yields is obtained.

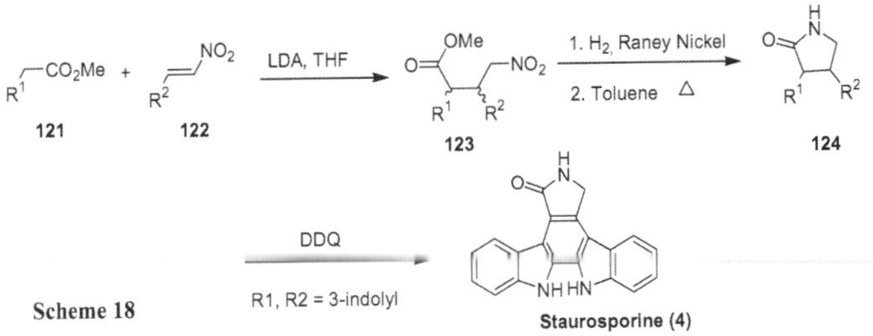

Scheme 18

3.4. Syntheses of the indolo[2,3-a]carbazole nucleus

Synthetic strategies for preparing the indolo[2,3-a]carbazole nucleus have been already summarized in Figure 2 based on the key bond formations, type of structure synthesized (aglycon), and research group. In the following section some of the methodologies are described briefly.

3.4.1. *Winterfeld's strategy to synthesis of staurosporinone*

In 1983, Winterfeld published the first synthesis of K252c as shown in Scheme 19 [52-53]. The synthesis of lactam **126** was successfully achieved by intramolecular aldol reaction of ketoamide **125** and then followed by titanium-mediated deoxygenation. Oxidative Photocyclization of **126** resulted in indolocarbazole **4** (staurosporinone).

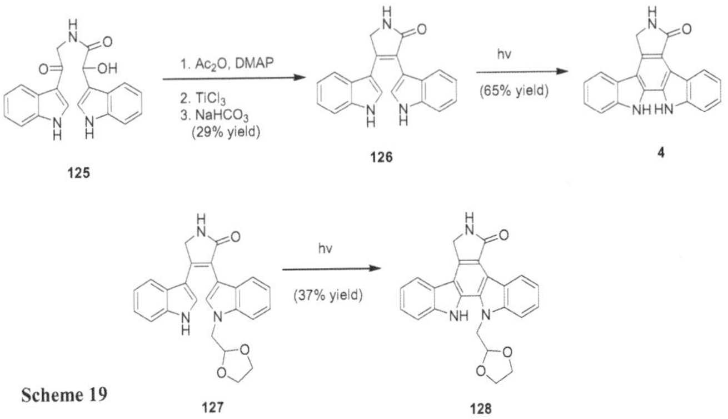

Scheme 19

3.4.2. *Magnus' approach*

Magnus published a synthetic methodology to selectively protect staurosporinones, just after Winterfeld's report [54]. Intramolecular Diels-Alder cycloaddition of indole-2,3-quinidomethane **130a** was the crucial step in his synthetic strategy (see Scheme 22). Imine **130** was prepared from condensation of tryptamine derivative **129** and 2-aminostyrene and then subjecting to acylation yielded indole-2,3-quinidomethane **130a** (*in situ*) and initiated an intramolecular Diels-Alder reaction. Oxidative work-up with DDQ resulted in indolocarbazole **131**. Deprotecting phthalimide group on **131** followed by acylation gave bis-protected staurosporinone **132**. Interestingly, the indoles could be selectively deprotected (e.g., **132→4** or **132→133**, Scheme 20) to facilitate regioselective introduction of a sugar portion.

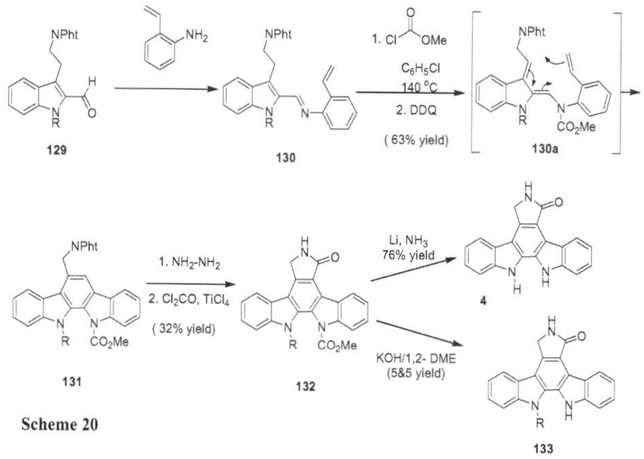

Scheme 20

3.4.3. *The Weinreb approach*

Weinreb exploited a synthetic strategy for the synthesis of bis indolyl maleimides to furnish maleimide **135** from indole-Grignard **134** and imide **134a** [43]. DDQ mediated oxidative cyclization of **135** resulted *N*-benzyl imide **136**. To complete the synthesis, Clemmenson reduction was performed for desymmetrizing **136**, to produce the corresponding lactam **137** (Scheme 21).

3.4.4. *Raphael's approach*

Raphael staurosporinone synthesis based on intermolecular Diels-Alder methodology and nitrene insertion chemistry is depicted in Scheme 22 [55-56]. Reaction of numerous dienophiles with diene **139** following dehydrogenation afforded triaryl products such as **140a** and **b**. In an initial attempt, **140b** was reduced and cyclized in good yield to afford lactam **137**, a compound previously prepared by Weinreb and Bergman [43].

3.4.5. *The Moody approach*

Moody utilized the pyranoindolone **147** to regulate intramolecular Diels-Alder reaction with subsequent aromatization to carbazole **148** (Scheme 23). Nitrene formation by deoxygenation using triethylphosphire produced K252c (**4**, staurosporinone) [57-58].

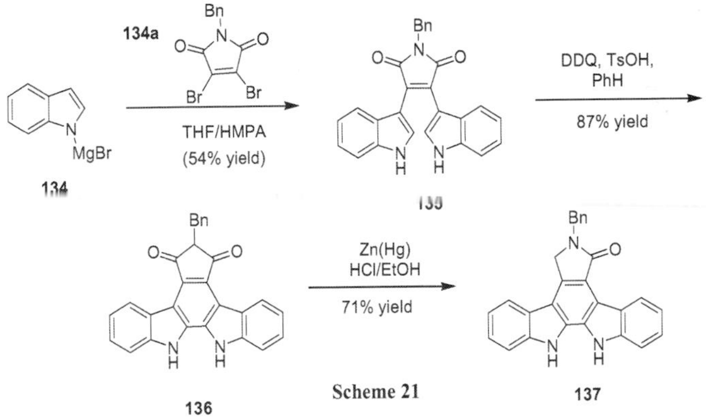

Scheme 21

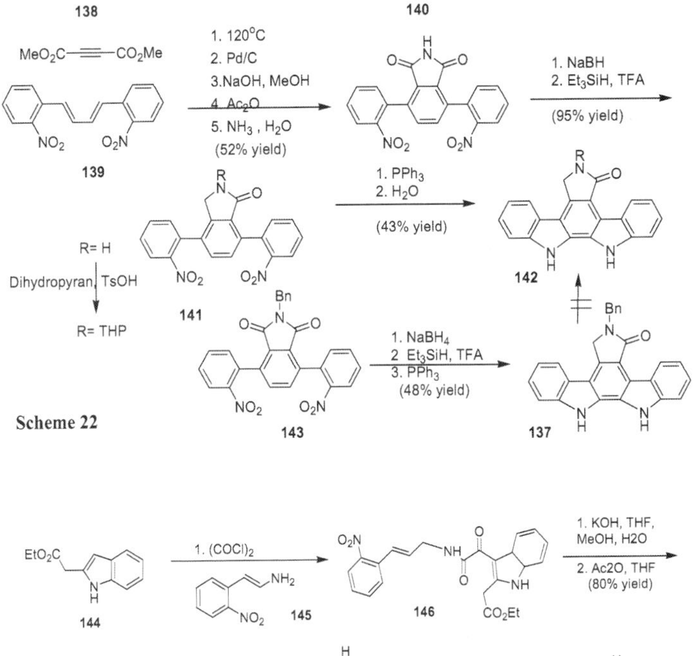

Scheme 22

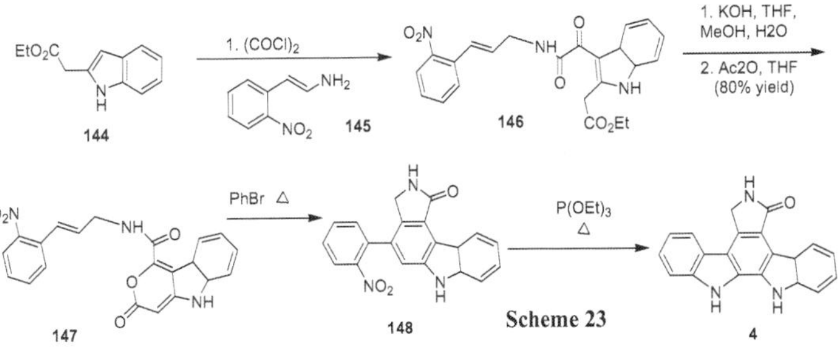

Scheme 23

4. Conclusion

In this book chapter, a brief introduction to biologically active indolocarbazole alkaloids was presented, with emphasis on the isolation and synthetic pathways of powerful protein kinase inhibitors such as Staurosporine indolocarbazole alkaloid and its analogues. Glycosylation on indolic moiety and concerns were discussed apart from the synthesis of staurosporinone aglycon and sugar portion. We do hope that this book chapter will be a valuable addition to the chemists dealing with indolocarbazole alkaloids from pharmaceutical industry and synthetic organic point of view.

Acknowledgements

Dr. Ravi Varala heartfully thanks Prof. Sirasani Satyanarayana, Vice-Chancellor (RGUKT-Basar) and Prof. Appala Naidu, RGUKT-AP, for their kind cooperation and support. And also, specially thank FAPESP-Brazil for the award of 'Visiting Researcher' grant (2014/25784-7). Profound thanks to Prof. S. J. Danishefsky, Prof. B. M. Stolz and Prof. J. L. Wood et al. for their valuable contributions to the field.

Author details

B. Purna Chandra Rao[1], Osvaldo N. Oliveira Jr.[2] and Ravi Varala[2*]

*Address all correspondence to:

1 Chemistry Department, Gitam University-Hyderabad Campus, Telengana, India

2 São Carlos Institute of Physics, University of São Paulo, São Carlos, SP, Brazil

References

[1] Shewach DS. Introduction to cancer chemotherapeutics. Chem. Rev. 2009; 109: 2859-2661.

[2] Cragg GM, Newman DJ. Plants as a source of anti-cancer agents. J. Ethnopharmacol. 2005; 100: 72-79.

[3] Cragg GM, Grothaus PG, Newman DJ. Impact of natural products on developing new anti-cancer agents. Chem. Rev. 2009; 109: 3012-3043.

[4] Sreedhar AS, Csermely P. Heat shock proteins in the regulation of apoptosis: new strategies in tumor therapy: A comprehensive review. Pharmacol. Ther. 2004; 101: 227-257.

[5] Pechan PM. Heat shock proteins and cell proliferation. FEBS Lett. 1991; 280: 1-4.

[6] McCubrey JA, Steelman LS, Abrams SL, Lee JT, Chang F, Bertrand FE, Navolanic P M, Terrian DM, Franklin RA, D'Assoro, AB, Salisbury JL, Mazzarino MC, Stivala F, Libra M. Roles of the RAF/MEK/ERK and PI3K/PTEN/AKT pathways in malignant transformation and drug resistance. Adv. Enzyme Regul. 2006; 46: 249-279.

[7] Lloyd DG, Golfis G, Knox AJ, Fayne D, Meegan MJ, Oprea TI. Oncology exploration: charting cancer medicinal chemistry space. Drug Discovery Today. 2006; 11: 149-159.

[8] Koehn FE, Carter GT. The evolving role of natural products in drug discovery. Nat. Rev. Drug Discovery 2005; 4: 206-220.

[9] Newman DJ, Cragg GM, Snader KM. Natural products as sources of new drugs over the period 1981-2002. J. Nat. Prod. 2003; 66: 1022-1037.

[10] Newman DJ. Natural Products as leads to potential drugs: An old process or the new hope for drug discovery?. J. Med. Chem. 2008; 51: 2589-2599.

[11] Akinaga S, Sugiyama K, Akiyama T. UCN-01 (7-hydroxystaurosporine) and other indolocarbazole compounds: a new generation of anti-cancer agents for the new century?. Anti-Cancer Drug Design. 2000; 15: 43-52.

[12] Prudhomme M. Rebeccamycin analogues as anti-cancer agents. Eur. J. Med. Chem. 2003; 38: 123-140.

[13] Mucke HA. CEP-1347 (Cephalon). IDrugs. 2003; 6: 377-383.

[14] Marminon C, Anizon F, Moreau P, Le´once S, Pierre A, Pfeiffer B, Renard P, Prudhomme M. Syntheses and antiproliferative activities of new rebeccamycin derivatives with the sugar unit linked to both indole nitrogens. J. Med. Chem. 2002; 45: 1330-1339.

[15] Knolker H-J, Reddy KR. Isolation and synthesis of biologically active carbazole alkaloids. Chem. Rev. 2002; 102: 4303-4427.

[16] Omura S, Iwai Y, Hirano A, Nakagawa A, Awaya J, Tsuchia H, Takakashi Y, Masuma R. A new alkaloid am-2282 of streptomyces origin taxonomy, fermentation, isolation and preliminary characterization. J. Antibiotics. 1977; 30: 275-282.

[17] Furusaki A, Hoshiba N, Matsumoto T, Hirano A, Iwai Y, Omura S. X-ray crystal structure of staurosporine: a new alkaloid from a streptomyces strain. J. Chem. Soc. Chem. Commun. 1978; 800-801.

[18] Furusaki A, Hoshiba N, Matsumoto T, Hirano A, Iwai Y, Omura S. The crystal and molecular structure of staurosporine, a new alkaloid from a *streptomyces* strain. Bull. Chem. Soc. Jpn. 1982; 5: 3681-3685.

[19] Gribble G, Berthel S. Studies in Natural Products Chemistry; Elsevier Science Publishers: New York, 1993; 12: 365-409.

[20] Sapi J, Massiot G. *The Alkaloids*; Academic Press: New York, 1995; 47: 173-226.

[21] Funato N, Takayanagi H, Konda Y, Toda Y, Hariyage Y, Iwai Y, Omura S. Absolute configuration of staurosporine by X-ray analysis. Tetrahedron Lett. 1994; 35: 1251-1254.

[22] Gani OA, Engh RA. Protein kinase inhibition of clinically important staurosporine analogues. Nat. Prod. Rep. 2010; 27(4): 489-498.

[23] Alessi DR. The protein kinase C inhibitors RO 31 8220 and GF 109203X are equally potent inhibitors of MAPKAP kinase -1b (Rsk-2) and p70 S6 kinase. FEBS Lett. 1997; 402: 121-123.

[24] Tamoki T, Nomoto H, Takahishi I, Kato Y, Morimoto M, Tomita F. Staurosporine, a potent inhibitor of phospholipid Ca±2 dependent protein kinase. Biochem. Biophys. Res. Commun. 1986; 135: 397-402.

[25] Prudhomme M. Biological targets of antitumor indolocarbazoles bearing a sugar moiety. Curr. Med. Chem.2004; 4: 509-521.

[26] Omura S, Iwai Y, Hirano A, Japan Kokai 78 73, 501; *Chem. Abstr.* 1978; 89: 178086b; idem, Ger. Offen. 2,745,326; *Chem. Abstr.* 1978;, 89: 58348y.

[27] Oka S, Kodama M, Takada H, Tomizuka N, Suzuki H, Agric G. Staurosporine, a potent platelet aggregation inhibitor from a *streptomyces* species. Biol. Chem. 1986; 50: 2723-2727.

[28] McAlpine JB, Karwoski JP, Jackson M, Mullaly MM, Hochlowski JE, Premachandran U, Burres NS. MLR-52, (4'-demethylamino4'5'-dihydroxystaurosporine), a new inhibitor of protein kinase c with immunosuppressive activity. J. Antibiotics 1994; 47: 281-288.

[29] Sato W, Yusa K, Naito M, Tsuruo T. Staurosporine, a potent inhibitor of C-kinase, enhances drug accumulation in multidrug-resistant cells. Biochem. Biophys. Res. Commun. 1990; 173: 1252-1257.

[30] Wakusawa S, Inoko K, Miyamoto K, Kajita S, Hasegawa K, Koyama M. Staurosporine derivatives reverse multidrug resistance without correlation with their protein kinase inhibitory activities. J. Antibiotics 1993; 46: 353-355.

[31] For a review, see: Bergman, J. Studies in Natural Products Chemistry; Elsevier Science Publishers: New York, 1988; 1: 3-30.

[32] Wood JL, Stoltz, B. M.; Dietrich, H.-J. Total Synthesis of (+)- and (-)-K252a. J. Am. Chem. Soc. 1995; 117: 10413-10414.

[33] Meksuriyen D, Cordell GA. Biosynthesis of Staurosporine, 1. 1H- and 13C-NMR assignments. J. Nat. Prod. 1988; 51: 884-892.

[34] Meksuriyen D, Cordell GA. Biosynthesis of staurosporine, 2. Incorporation of tryptophan. J. Nat. Prod. 1988; 51: 893-899.

[35] Pearce CP, Doyle TW, Forenza S, Lam KS, Schroeder DR. The biosynthetic origins of rebeccamycin. J. Nat. Prod. 1988; 51: 937-940.

[36] Yang S-W, Cordell GA. Biosynthesis of staurosporine: Incorporation of glucose. J. Nat. Prod. 1996; 59: 828-833.

[37] Hoehn P, Ghisalba O, Moerker T, Peter HH. "3'-Demethoxy-3'-hydroxystaurosporine, a novel staurosporine analogue produced by a blocked mutant." J. Antibiot. (Tokyo) 1995; 48(4): 300-305.

[38] Weidner S, Kittelmann M, Goeke K, Ghisalba O, Zahner H. 3'-Demethoxy-3'-hydroxystaurosporine-O-methyltransferase from *Streptomyces longisporoflavus* catalyzing the last step in the biosynthesis of staurosporine. J. Antibiot. (Tokyo) 1998; 51(7): 679-682.

[39] Howard-Jones AR, Walsh CT. Staurosporine and rebeccamycin aglycones are assembled by the oxidative action of StaP, StaC, and RebC on chromopyrrolic acid. J. Am. Chem. Soc. 2006; 128 (37): 12289-12298.

[40] Link JT, Raghavan S, Danishefsky SJ. First total synthesis of staurosporine and *ent*-Staurosporine. J. Am. Chem. Soc. 1995; 117: 552-553.

[41] Link JT, Gallant M, Danishefsky SJ. The first synthesis of a fully functionalized core structure of staurosporine: sequential indolyl glycosidation by endo and exo glycols. J. Am. Chem. Soc. 1993; 115: 3782-3783.

[42] Winterfield E. In Heterocycles in Bio-Organic Chemistry; Bergmann, J., Ed.; The Royal Society of Chemistry: London, 1991; 18-27.

[43] Joyse RP, Gainor JA, Weinreb SM. Synthesis of the aromatic and monosaccharide moieties of staurosporine J. Org. Chem. 1987; 52: 1177-1185.

[44] McCombie SW, Bishop RW, Carr D, Dobek E, Kirkup MP, Kirschmeir P, Lin SI, Petrin J, Rosinski K, Shankar BB, Wilson O. Indolocarbazoles. 1. Total synthesis and protein kinase inhibiting characteristics of compounds related to K-252c. Bioorg. Med. Chem. Lett. 1993; 3:1537-1542.

[45] Link JT, Raghavan S, Gallant M, Danishefsk SJ, Chou TC, Ballas LM. Staurosporine and *ent*-Staurosporine: The first total syntheses, prospects for a regioselective approach, and activity profiles J. Am. Chem. Soc. 1996; 118, 2825-2842.

[46] Wood JL, Stolz BM, Goodman SN, Onwuene K. Design and implementation of an effective synthetic approach to pyranosylated indolocarbazoles: Total synthesis of (+)-RK286c, (+)-MLR-52, (+)-Staurosporine, and (-)-TAN-1030a. J. Am. Chem. Soc. 1997; 119: 9652-9661.

[47] Weinreb SM, Garigipati RS, Gainor JA. Natural product synthesis *via* cycloadditions with *N*-sulfinyl dienophiles. Heterocycles. 1984: 21; 309-324.

[48] Gallant M, Link JT, Danishefsky SJ. A Stereoselective synthesis of indole-8-*N*-glycosides: An application to the synthesis of Rebeccamycin J. Org. Chem. 1993; 58: 343-349.

[49] Anizon F, Moreau P, Sancelme M, Voldoire A, Prudhomme M, Ollier M, SeveÁre D, Riou J-F, Bailly C, Fabbro D, Meyer T, Aubertin AM. Syntheses, biochemical and biological evaluation of staurosporine analogues from the microbial metabolite rebeccamycin. Bioorg. Med. Chem. 1998; 6: 1597-1604.

[50] Rajeshwaran GG, Mohanakrishnan AK. Synthetic studies on indolocarbazoles: Total synthesis of staurosporine aglycon. Org. Lett. 2011; 13 (6): 1418-1421.

[51] Mahboobi S, Eibler E, Koller M, Kumar KCS, Popp A. Synthesis of pyrrolidin-2-ones and of staurosporine aglycon (K-252*c*) by intermolecular Michael reaction. J. Org. Chem. 1999; 64: 4697- 4704.

[52] Burkhard S, Winterfeld E. Reactions with indole derivatives, XLVIII. A simple synthesis of the staurosporine aglycon. Heterocycles 1983; 20: 469-476.

[53] Bruning J, Hache T, Winterfeld E. The First regioselective synthesis of a monosubstituted staurosporine precursor. Synthesis. 1994; 25-27.

[54] [Magnus PD, Sear NL. Indole-2,3-quinodimethanes : Synthesis of selectively protected derivatives of the fused dimeric indole alkaloid staurosporinone. Tetrahedron. 1984; 40: 2795-2797.

[55] Hughes I, Raphael RA. Synthesis of arcyriaflavin B. Tetrahedron Lett. 1983; 24: 1441-1444.

[56] Hughes I, Nolan WP, Raphael RA. Synthesis of the indolo[2,3-*a*]carbazole natural products staurosporinone and arcyriaflavin B. J. Chem. Soc., Perkin Trans. 1. 1990; 2475-2480.

[57] [57] Moody CJ, Rahimtoola KF. Synthesis of the staurosporine aglycone. J. Chem. Soc., Chem. Commun. 1990; 1667-1668.

[58] Moody CJ, Rahimtoola KF, Porter B, Ross BC. Synthesis of the staurosporine aglycon. J. Org. Chem. 1992; 2105-2114.

[59] For a comprehensive review on the therapeutic potential of PKC inhibitors, see: Bradshaw, D.; Hill, C. H.; Nixon, J. S.; Wilkinson, S. E. Agents and Actions. 1993; 38: 135-147.

Chapter 5

Breakthroughs in Indole and Indolizine Chemistry – New Synthetic Pathways, New Applications

Ioana Otilia Ghinea and Rodica Mihaela Dinica

Additional information is available at the end of the chapter

http://dx.doi.org/10.5772/62079

Abstract

Indole and indolizines (heterocyclic aromatic compounds structurally and chemically isomeric with indoles) are an important class of N-fused heterocyclic compounds due to their interesting biological and optical properties. Different strategies for generating diverse collections of small molecules with indole and indolizine moieties have been developed. They can be synthesized by means of classical and nonclassical pathways. The present study discusses the versatile nature of indole/indolizine derivatives, new green methods for their synthesis, their possible mechanism of action and also provides information about current/future prospects of the topics and different indole/indolizine derivatives in pharmaceutical/clinical trials. With the remarkable number of approved indole-containing drugs as well as the importance of the indolizine moiety, it can be easily concluded that indole and indolizine derivatives offer perspectives on how pyrrole scaffolds might be exploited in the future as bioactive molecules against a broad range of diseases.

Keywords: Indole, indolizine, bioactive heterocycles, green chemistry, functionalization, mechanism

1. Introduction

A great deal of research in heterocyclic chemistry concerns the development of strategies for efficient synthesis and the discovery of new methods of ring formation, since more than half of the biologically active compounds produced by nature contain a heterocyclic moiety as a fundamental unit in their structure. Also, heteroaromatic compounds are always of great importance for chemists and the identification and confirmation of highly potent and selective bioactive molecules is a decisive step both in academic and pharmaceutical research.

Heterocyclic compounds with a pyrrole cycle are significant both in materials and in medicinal chemistry [1]. Indoles and indolizines (heterocyclic aromatic compounds structurally and chemically isomeric with indoles) are important classes of N-fused heterocyclic compounds due to their interesting biological and optical properties. Although their chemistry is a well-established subject for researchers, they continue to attract much attention due to their diverse biological properties. Also, the correlation between indoles and indolizines has prompted speculation that indolizine analogs of biologically important indoles could conceivably have potent physiological activities [2].

Indoles and their derivatives are well-known as an important class of heterocyclic compounds, their core being a near-ubiquitous component of biologically active natural products, widespread in different species of plants, animals, and marine organisms. The indole is also well-known as one of the most important scaffolds for drug discovery, capable of serving as ligand for a diverse array of receptors and it has been a major focus of research [3]. Indole derivatives have the unique property of mimicking the structure of peptides and to bind reversibly to enzymes and exhibit significant physiological and pharmacological, industrial, and synthetic applications such as beneficial estrogen metabolism promoter in humans, anticarcinogenic properties, inhibitors of human prostate cancer cells, and free radical scavenging activities [1, 4]. The indole scaffold is widely used in antiviral drugs and reverse-transcriptase inhibitors, drugs used to treat HIV infection or AIDS, and in some cases hepatitis B. Meanwhile, a number of bis (indolyl) alkanes have received considerable attention because of their occurrence in bioactive metabolites of terrestrial and marine origin [5].

Indolizine is the core structure of many of the naturally occurring alkaloids such as swainsonine (a potent inhibitor of Golgi alpha-mannosidase II, an immunomodulator and a potential chemotherapy drug), monomorine (might be used to lure ants to their doom), gephyrotoxin (muscarinic antagonist), and lamellarins (HIV-1 integrase inhibition and antibiotic activity) [6].

The indolizine synthetic derivatives also deserve special attention because of their pharmacological properties such as antibacterial, anti-inflammatory, antiviral and antileishmanial, analgesic and antitumor, antioxidant activities, aromatase inhibition, calcium entry blocking, histamine H3 receptor antagonist, and physicochemical properties such as strong fluorescence [6, 7].

Different strategies for generating diverse collections of small molecules with indole and indolizine moieties have been developed. They can be synthesized by means of classical and nonclassical pathways.

The development of simple, convenient, and an eco-friendly approach for the synthesis of these biologically important compounds is still in demand. For example, the very useful and green concept of a "click" reaction is a facile, selective, high-yield reaction under mild water-tolerant conditions with little or no by-products [8]. Cascade annulation reactions lead also to the formation of polycyclic fused six- and seven-membered heterocycles with indole and indolizine core [9].

Microwave irradiation, sonication, and solvent-free are green chemistry techniques that have been used for a variety of applications including organic synthesis. Microwaves and ultra-

http://dx.doi.org/10.5772/62079

sounds have been used as synthetic techniques for obtaining indole and indolizine derivatives in high-yield, higher reaction rate. The simplicity of the reactions using these techniques, the elimination of toxic solvents, and the synthesis carried out in a very short time period are particularly useful for the creation of diverse chemical compounds of "drug-like" molecules for biological screening [10].

Multicomponent reactions (MCRs) or tandem reactions have developed as a powerful tool for delivering the molecular diversity needed for the synthesis of interesting heterocyclic scaffolds, to efficiently construct a variety of intermediates possessing an indolyl or indolizyl subunit and are particularly attractive especially if they start from simple molecules [11].

The present study discusses the versatile nature of indole/ indolizine derivatives, new green methods for their synthesis, their possible mechanism of action, and also provides information about current/future prospects of the topics and different indole/indolizine derivatives in pharmaceutical/clinical trials.

2. Indoles

Indole derivatives are, perhaps, the most studied nitrogen heterocyclic systems because of interesting biological properties that received particular interest due to the reserpine alkaloid, one of the first drugs used for the treatment of central nervous system (CNS) disorders. Different substituted indoles are particularly important in pharmaceutical chemistry being capable to bind many receptors with high affinity exhibiting various pharmacological activities. Therefore, it is important to explore new synthetic reactions and evaluate various properties of indole derivatives.

2.1. Indole synthesis

To obtain biologically relevant N-hydroxyindoles, a prudent step would be to synthesize O-protected hydroxyindoles, to avoid their dimerization into kabutanes. Such were the premises of one study, presenting the annulation of nitrosoarenes with various alkylating and acylating agents, able to afford the desired compounds with excellent regioselectivity [4].

The synthesis of 3,3-dimethyl-2-amide indoles could be achieved through the I2/DMSO promoted oxidative amidation reaction between 1,2,3,3-tetramethyl-3H-indolium iodide and secondary amines with moderate yields (Figure 1) [12].

Using a method involving four steps, 2-indole-3-yl-thiochroman-4-ones could be obtained (Figure 2), according to Song et al. In the final step, the Michael addition reaction of thiochromone and indole, an ionic liquid is used, to increase the yield, with the added advantage that it could be reused three times without a decrease of efficiency [13].

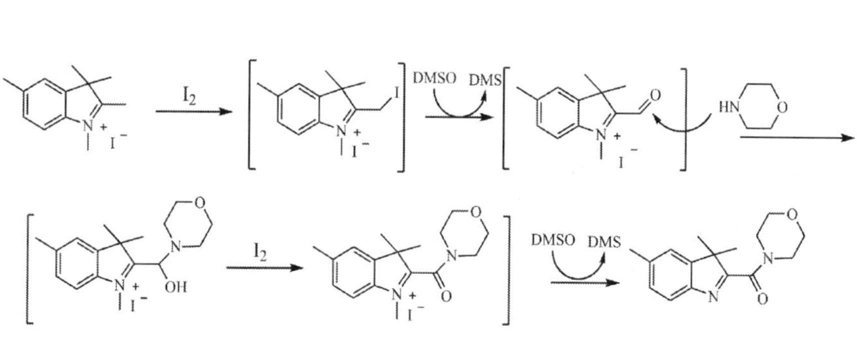

Figure 1. Synthesis of 3,3-dimethyl-2-amide indoles: mechanism [12]

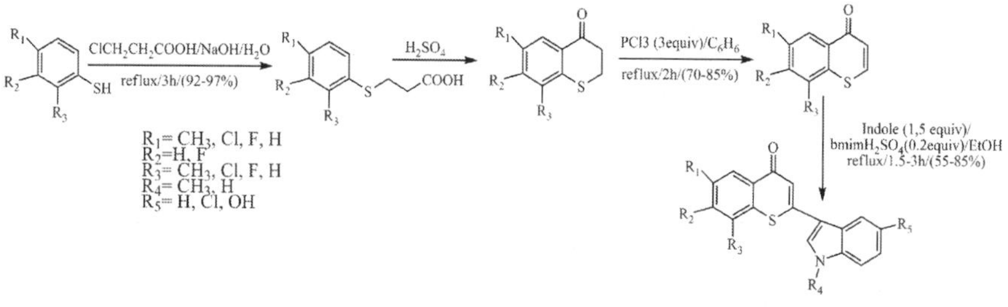

Figure 2. Synthesis of 2-indole-3-yl-thiocroman-4-ones derivatives [13]

2.2. Green methods for indole synthesis

Polyvinylsulfonic acid, a biodegradable and recyclable polymeric acid rarely used in organic transformations, could be used as a Bronsted acid catalyst in the synthesis of bis (indolyl) methane [14]. Another pathway to obtain this compound would be to employ a reusable resin, Indion Ina 225H, as catalyst of the substitution reaction between indoles and aldehydes (Figure 3), reportedly attaining excellent yields in short reaction times [15].

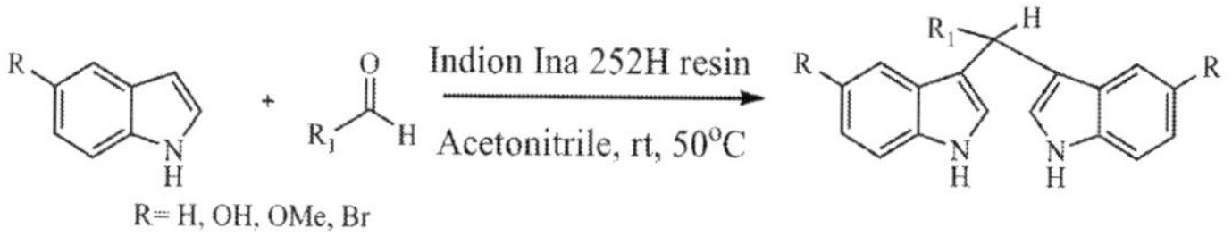

Figure 3. Synthesis of bis(indolyl) methane using a reusable resin, Indion Ina 225H [15]

Various carbonyl compounds, including ketones could also be building blocks for the much desired bis(indolyl)methanes, using catalytic amounts of iodine in the presence of sodium dodecylsulfate in aqueous solution above its critical micellar concentration and the protocol was also extended to afford 3-substituted indolyl ketones (Figures 4 and 5) [3].

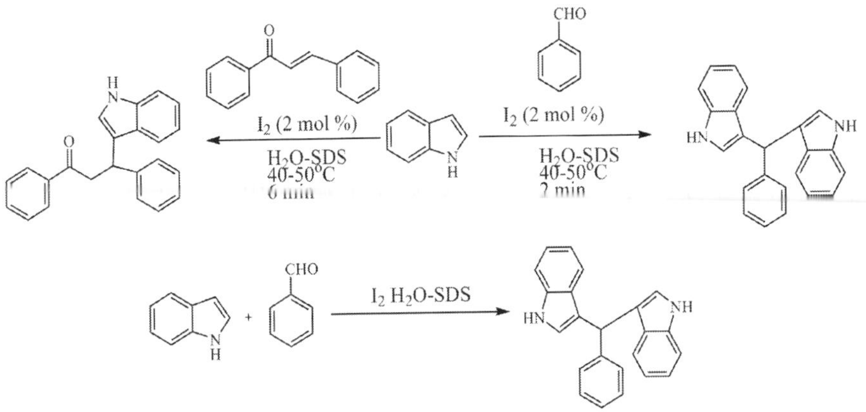

Figure 4. Synthesis of bis(indolyl)methanes, using catalytic amounts of iodine in the presence of sodium dodecylsulfate [3]

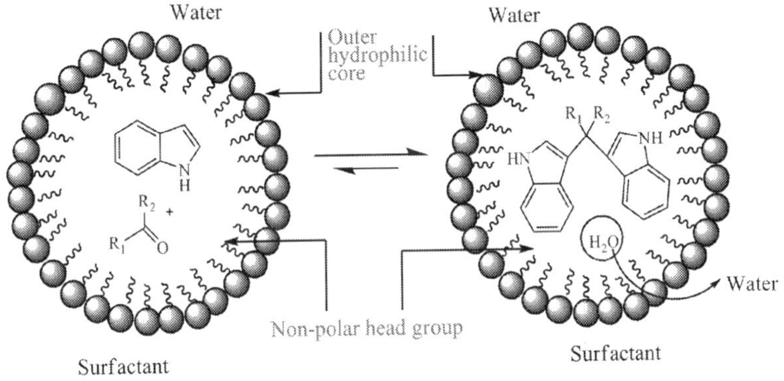

Figure 5. Iodine-catalyzed formation of bis(indolyl) methane from indole and aldehyde under aqueous micelar conditions [3]

3. Indolizine derivatives

The indolizine core has found numerous applications in the synthesis of biologically active compounds. Partially or completely reduced indolizine analogs are widely used in the synthesis of indolizidine alkaloids and related unnatural products. Among many other pharmacological uses, polycyclic analogs of indolizine, for example, have found a broad application as heterocyclic analogs of indene in the synthesis of ligands for transition metal complexes. Although many methods have been developed for their synthesis, they are generally time-consuming or require the use of stoichiometric amounts of organometallic reagents, Lewis acids, expensive catalysts or potentially toxic solvents, which limits their economical applications. Accordingly, development of alternative catalytic methods for construction of these important heterocyclic cores is necessary [9].

3.1. Indolizines obtained via 1,3-Dipolar cycloaddition

The mechanism of obtaining condensed five-membered ring systems via 1,3-dipolar cycloadditions implies the reaction of a dipole, in this case an N-ylide generated in situ from a cycloimmonium halide and a base or another deprotonating agent followed by its addition to a dipolarophile, olefinic, or acetylenic [7].

N-ylides could be generated employing 1,2-epoxybutane as both solvent and deprotonation agent, or by using triethylamine in DCM, with ethyl propiolate or DMAD as dipolarophiles, or again coupling the ylide with acrylonitrile and using TPCD for the aromatization step, all methods with similar medium to good yields (Figure 6) [7]. Moderate yields, up to 22%, are reported when reacting DMAD with N-ylides generated from pyridinium salts and K_2CO_3, using catalytic amounts of dicyclohexyl-18-crown-6 [16]. In another study, 20 substituted indolizines were obtained in just 30 min at room temperature, employing electron-deficient alkynes, in the presence of K_2CO_3 in DMF, with yields as high as 77% [17].

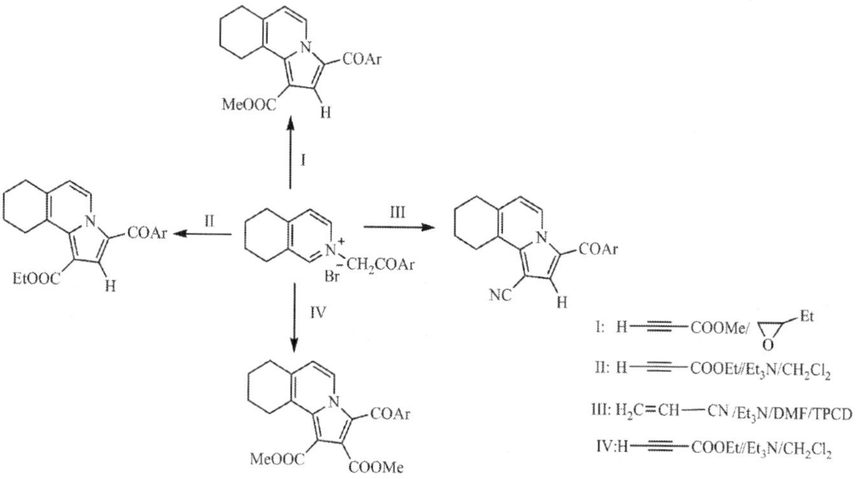

Figure 6. Synthesis of the new 7,8,9,10-tetrahydropyrrolo[2,1-a]isoquinolines [7]

The solvent could have a great impact on the reaction mechanism, as one study demonstrates, using substituted ethenes (E-1,2-di(alkylsulfonyl)-1,2-dichloroethene) as dipolarophiles (Figure 7). In aprotic solvents, the reaction takes place as a 1,3-dipolar cycloaddition, with

http://dx.doi.org/10.5772/62079

yields between 62 and 75% for the six indolizines obtained, but in protic solvents an addition–elimination reaction intervenes, leading to the competitive formation of furans, with indolizine yields as low as 9% [18].

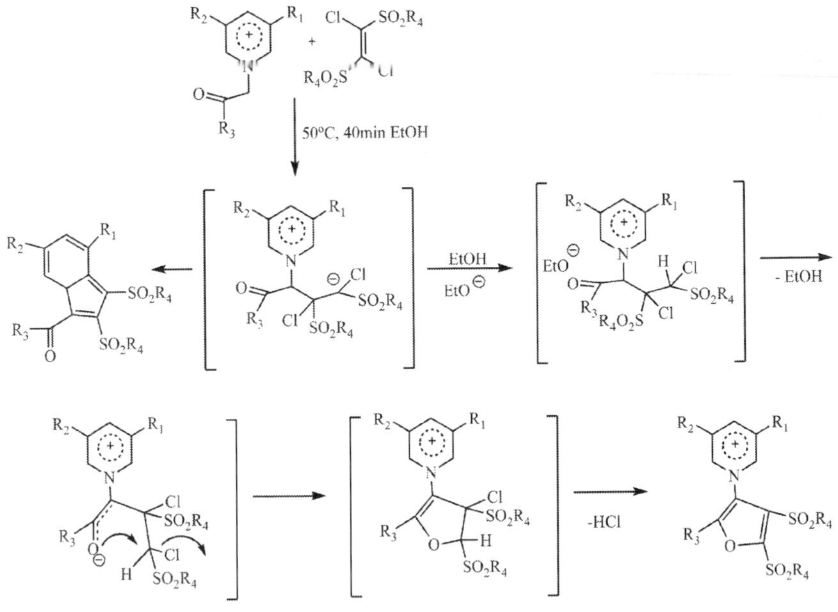

Figure 7. The reactions of pyridinium ylides with ethenes in EtOH [18]

3.2. Indolizines obtained via one-pot reactions

One-pot reactions imply obtaining the product in a single step, by adding all the necessary reagents in the same reaction medium, without having to isolate and purify any precursors of the desired product. This type of procedure offers advantages such as swiftness, the preparation of complex compounds from readily available material, simplification of workup and atom economy.

Mishra et al. present a method to obtain 1-aminoindolizines from aldehydes, secondary amines, and terminal alkynes, in a one-pot reaction (Figure 8). After testing several solvents and metal catalysts, the best results are obtained with CuCl in PEG, synthesizing 15 substituted indolizines with yields exceeding 70%, after 3–4 h reaction time [19].

Substituted 3-aminoindolizines could be obtained via one-pot multistep reactions, from 2-pyridine carboxyaldehide and various nitriles, after 3 h reaction in toluene at 105°C, by adding 1.1 eq of Hantzsch ester as a hydride transfer agent and catalytic amounts of piperidinium acetate [20].

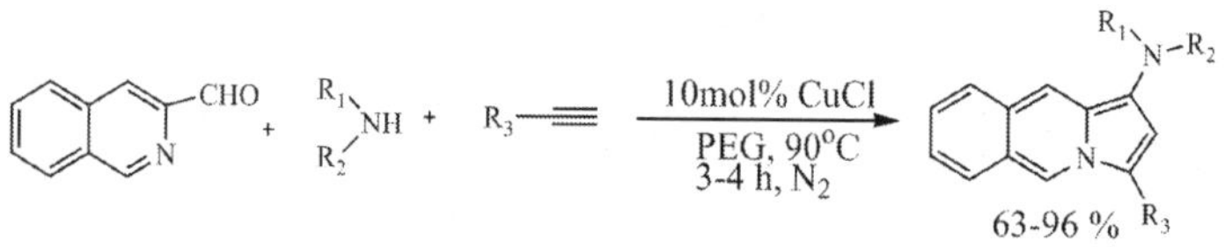

Figure 8. One-pot multicomponent synthesis of 1-aminoindolizines [19]

The synthesis of 2-acetyl-3-(phenylamino)indolizine-1-carboxamides could also be achieived in a single step (Figure 9), by combining pyridine-2-carbaldehyde, acetoacetanilide and isocyanides in toluene at reflux, with yields around 90% for the four compounds obtained [21].

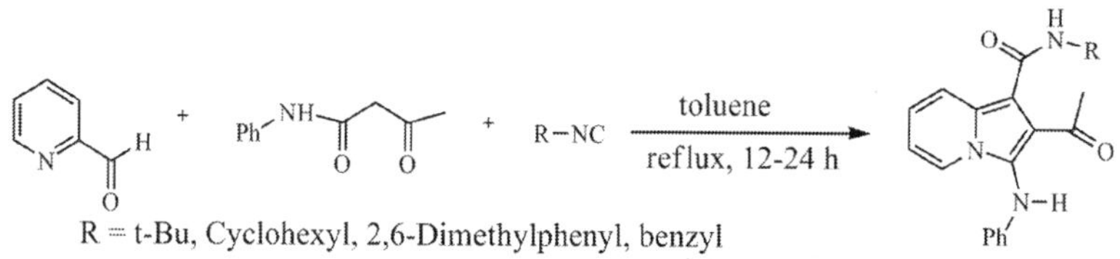

Figure 9. Synthesis of 2-acetyl-3-(phenylamino)indolizine-1-carboxamides via a three-component condensation [21]

A four-component tandem reaction is proposed by Zhenjun et al., by treating pyridine (or quinoline) with phenacyl bromides (or bromoacetophenones), ethyl glyoxalate, and Na_2CO_3 in refluxing acetonitrile. The resulting polysubstituted indolizines are obtained after 16 h of reaction time in moderate-to-good yields [22].

Seventeen polysubstituted indolizines could be obtained via a one-pot sequential addition-cyclodehydration-dehydrocyanation from of 2-(1H-pyrrol-1-yl) nitriles with α,β-unsaturated carbonyl compounds (Figure 10) [23].

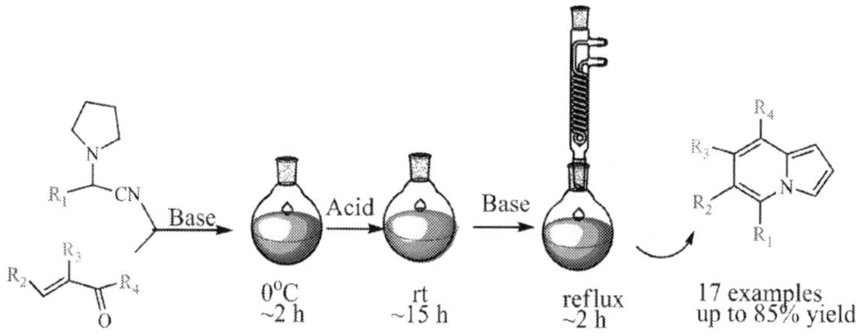

Figure 10. One-pot addition-cyclodehydration-dehydrocyanation of 2-(1H-pyrrol-1-yl) nitriles with α,β-unsaturated carbonyl compounds [23]

3.3. Novel approaches to indolizine synthesis

The indolizine core could be accessible starting from pyrrole, with strategies involving intramolecular aldol cyclization or domino Knoevenagel condensation, shown in Figure 11 [24].

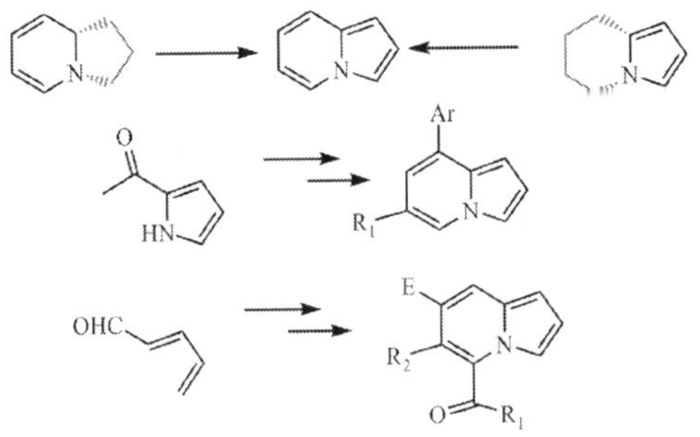

Figure 11. Possible synthetic approaches to indolizines [24]

Another [3+3] annulation approach employs allyl bromides derived from Morita-Baylis-Hilman adducts (Figure 12), with the conclusion that electron withdrawing groups, as substituents at the aromatic ring, contribute to successful ring closure and result in accordingly substituted indolizines [25].

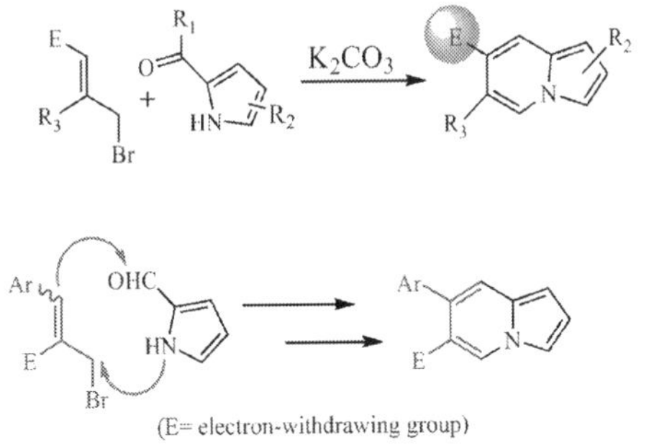

Figure 12. Design of new [3+3] annulation route to indolizines [25]

Novel 2-acyl-6-aryl substituted indolizines were obtained starting from 4-acyl-pyrrole-2-carbaldehyde and α, β-unsaturated esters, in the presence of K_2CO_3 in DMF, with yields between 42 and 68% after 8–12 h at 50°C [26].

Another possibility would be to perform a tandem oxidative C-H functionalization and 5-endo-dig cyclization, starting from 2-substituted pyridines and alkynes (Figure 13), which could be achieved with good yields using an Ag_2CO_3 reusable catalyst [27].

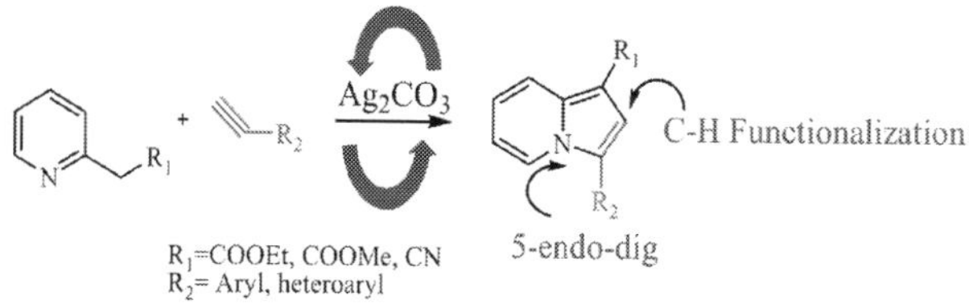

Figure 13. Silver-mediated oxidative C-H functionalization to synthesize indolizines [27]

Substituted pyridines and acetophenones lead to the formation of 1,2,3-triarylindolizines with moderate-to-excellent yields, promoted by I_2/DMSO at 100°C, the proposed mechanism for this reaction is presented in Figure 14 [28].

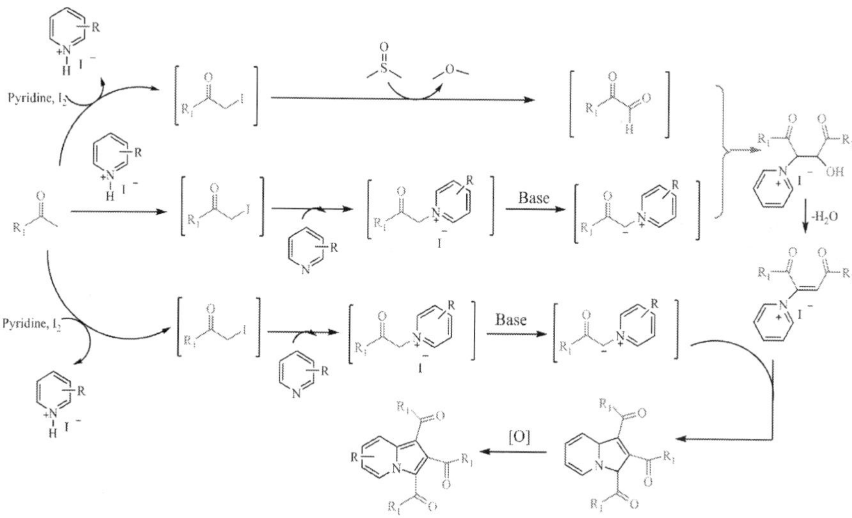

Figure 14. Formation of 1,2,3-triarylindolizines. Proposed reaction pathway [28]

3.4. Green methods for indolizine synthesis

During the synthesis of N-heterocycles there are many problems of health and safety in addition to the environmental problems caused by their use and disposition as waste. Green methods are a route towards increasing the efficiency of indoles and indolizines synthesis, and stride to use less toxic solvents, to reduce the stages of the synthetic routes and minimize waste as far as practically possible for sustainable development.

A potential method to make synthetic chemistry more environment-friendly would be to reuse catalysts, such as ion-exchanging resins. Amberlite-IRA 402 (OH) could be employed as the

http://dx.doi.org/10.5772/62079

ylide-forming base in the reaction between phenacyl pyridinium, quinolinium and isoquinolinium salts and alkynes (Figure 15) [29].

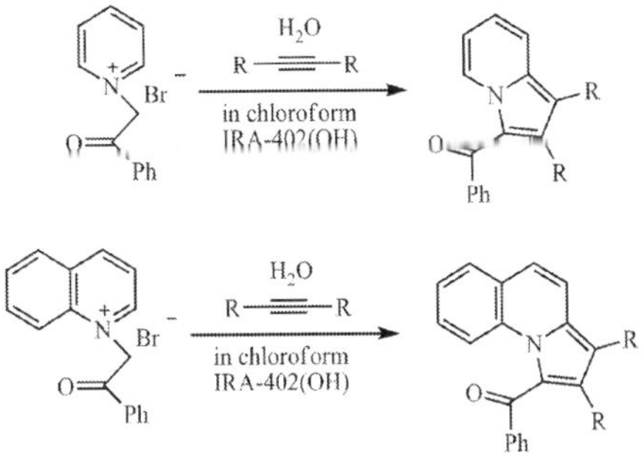

Figure 15. Synthesis of indolizines and pyrrolo [1,2-a] quinolines using alkynes [29]

Unconventional activation techniques, such as microwave irradiation, not only lead to shorter reaction times but generally increase the purity of the desired compound. One study presents the synthesis of 8 indolizine derivatives in an aqueous medium that were obtained in good yields after 1 min of irradiation at 300W [30].

Biocatalysis could be employed to aid the formation of indolizine derivatives in an aqueous medium, as seen in Figure 16. Ultrasound activation was compared to conventional heating, affording 7,7′-bis-indolizines with similar yields in much shorter reaction times [31].

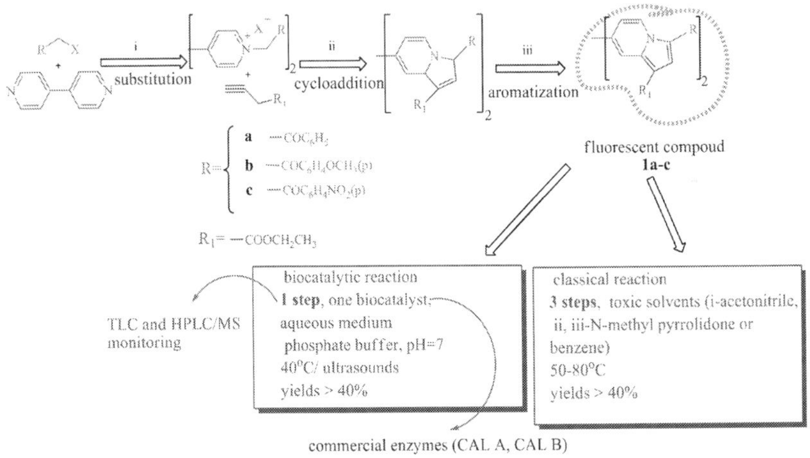

Figure 16. Synthesis of bisindolizines by biocatalytic reaction [31]

4. Indoles and indolizines functionalization

The Oxone-induced oxidation of indole-3-carbaldehydes and 5-halogenated analogs could lead to the formation of tryptanthrin derivatives (Figure 17), a highly functionalized biologically active natural product, at room temperature [32]. The phthalazine moiety could also be accessible with green methods, employing catalytic amounts of L-proline, with good yields and less than 2 h reaction time [33].

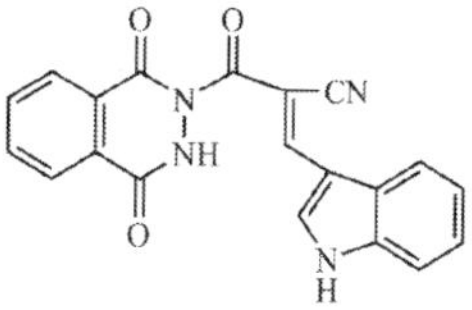

Figure 17. Tryptanthrin derivative [32]

The reaction between indole and formaldeyde could produce high yields of hemiaminals, with the added value of mild conditions, such as room temperature and an aqueous medium, in the presence of TBAF [34].

The most eco-friendly approach when it comes to solvent choices would be not to employ any solvents. Analogs of 3-alkylindole, for example, could be prepared in solvent-free conditions, using MgO nanoparticles as catalyst [35]. Bis(indol-3-yl)methanes could be synthesized in solvent-free grinding conditions, employing a reusable catalyst, phosphate-impregnated titania, obtaining yields as high as 93% [36].

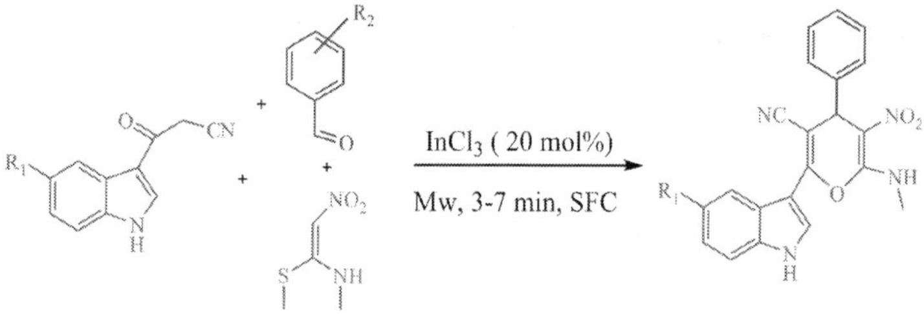

Figure 18. Synthesis of highly functionalized indolylpyrans [37]

Unconventional activation techniques could also be used for indole functionalization. Within 10 min, including reaction and purification time, 3-pyranyl indole derivatives could be obtained with good yields, through one-pot microwave-assisted reactions, with $InCl_3$ as catalyst (Figure 18) [37]. Indolyl chalcones could be prepared from indole-3-carboxaldehyde and heteroaryl active methyl compounds under conventional heating, but the yield was much improved and reaction time was drastically reduced (from more than 9 h to less than 15 min) when microwave irradiation was introduced [38]. Ultrasounds aid the selective formation of

http://dx.doi.org/10.5772/62079

11 3-selanylindole derivatives with good yields, proving superior to conventional heating and microwave irradiation for this synthesis [39].

Both indoles and indolizines could be functionalized via alkylation with enamides under mild conditions (Figures 19 and 20), using $FeCl_3$, in short reaction times with good yields [2].

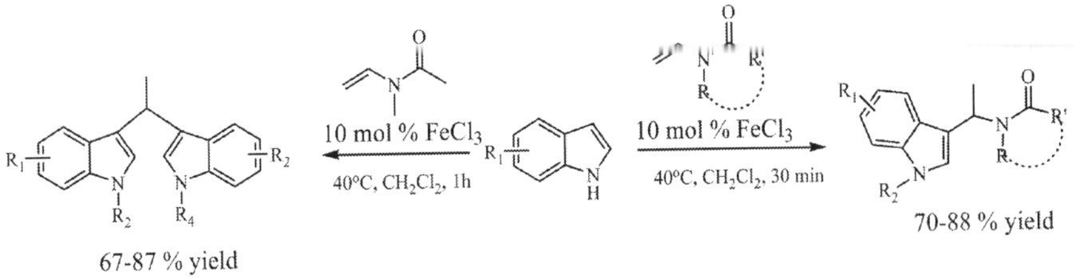

Figure 19. Iron-catalyzed alkylation of indoles with enamides [2]

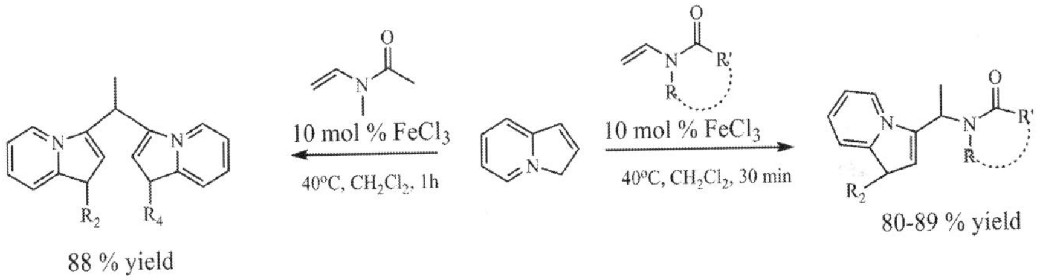

Figure 20. Iron-catalyzed alkylation of indolizines with enamides [2]

The Friedel-Crafts alkylation of indoles could also be performed in water, as presented in Figure 21, with yields as high as 97% in the presence of Keggin heteropoly acids, solid super-acid catalysts [40].

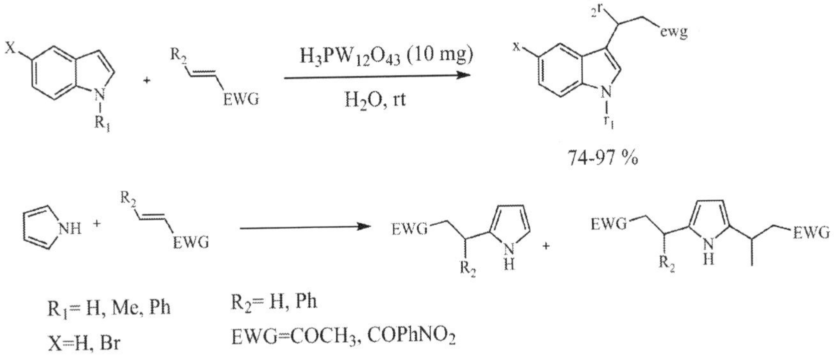

Figure 21. Friedel-Crafts alkylation of indole [40]

A novel approach presents the previously inaccessible regioselective formation of substituted pyrido[2,3-b]indolizine-10-carbonitriles, via a cascade transformation of α,β-unsaturated carbonyl compounds with a dimer of 1-(cyanomethyl) pyridinium chloride, in ethanol/water in the presence of sodium acetate [41].

5. Pharmaceutical applications

As we have seen so far, interesting pathways have been proposed for the synthesis of indoles and indolizines. Many of these molecules have subsequently been involved in tests to assess their biological activity. Natural compounds with these moieties have also attracted interest, not just as extracts, but as targets for total/semisynthesis or as frameworks for compound libraries. Next, we shall review some of the extremely diverse pharmaceutical applications of these derivatives, ranging from fluorescence probes, to antiviral, to anticancer molecules currently in clinical trials.

5.1. Natural and synthetic indoles

Lead by Cialis, there are seven indole-containing commercial drugs in the Top-200 Best Selling Drugs by US Retail Sales in 2012. Examples of indole derivatives marketed as antiviral drugs, for example, are Arbidol (a broad-spectrum antiviral with anti-influenza and immunomodulating effects) and Delavirine (a non-nucleoside reverse transcriptase inhibitor) [42].

Bisindoles, such as hamacanthin A, isolated from marine sponges (*Hamacantha* sp., *Spongosorites* sp.), or the more famous indole-3-carbinol (I3C), a compound found in cruciferous vegetables (cabbage, kale, cauliflower, broccoli, Brussels sprouts) and its bisindole metabolite, 3,3'-diindolylmethane (DIM), have displayed biological activities such as antimicrobial, antiparasitic, anti-inflammatory, and anticancer and are high up on the interest list of many researchers [43].

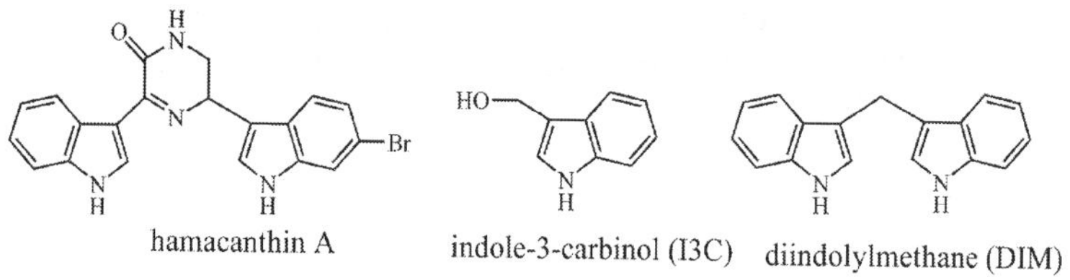

Figure 22. Natural indoles [50]

Some of the many studies published in this field have resulted in the elucidation of some of the mechanisms of their bioactivity. The influence of I3C, for example, on lung cancer cells, has been attributed to apoptosis via Fas activation and caspase-8 pathways and also cell-cycle arrest at the G0/G1 phase, and it was also shown that cancer preventive effects of I3C were mediated via modulation of the phosphatidylinositol-3-kinase (PI3K)/Akt signaling pathway

[44]. I3C was also shown to induce the expression of phase I and II enzymes by the binding of the aryl hydrocarbon receptor (AhR) (Figure 23) [45].

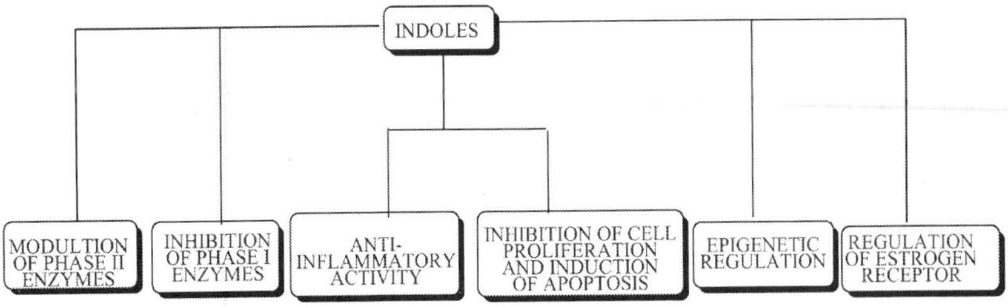

Figure 23. Biological activity pathways of indoles [45]

DIM has been found to increase bone mass by supressing osteoclastic bone resorption, in physiological and pathological conditions [46]. DIM could also help prevent heart failure, as one study indicates the compound improves myocardial energy metabolism imbalance via AMPKα signaling [47].

Strychnine and brucine are well known for their toxic effect that manifests in the form of hypertension and violent convulsions. Brucine is also a proposed anticancer drug candidate, as it inhibits VEGF-induced cell proliferation, reducing p-VEGFR2 kinase activity and inhibiting neovascularization in vivo [48].

Other indole alkaloids, isolated from marine sources, such as coscinamides, dragmacidin D, topsentins, or even fungal sources, such as asterriquinone, have exibited antiviral (anti-HIV), antimicrobial, antitumor activity, along with the inhibition of serine–threonine protein phosphatases or ascites hepatoma AH13, for example [49]. Such compounds, isolated from the *Strychnos* species, have also been found to inhibit quinine- and choloquinine-resistant *P. falciparum* [50]. One of the more studied indole alkaloids would be physostigmine, the template that led to the development of rivastigmine, globally licensed in 2006 to fight the symptoms of dementia associated with Parkinson's disease, also prescribed for the symptomatic treatment of Alzheimer's disease [51].

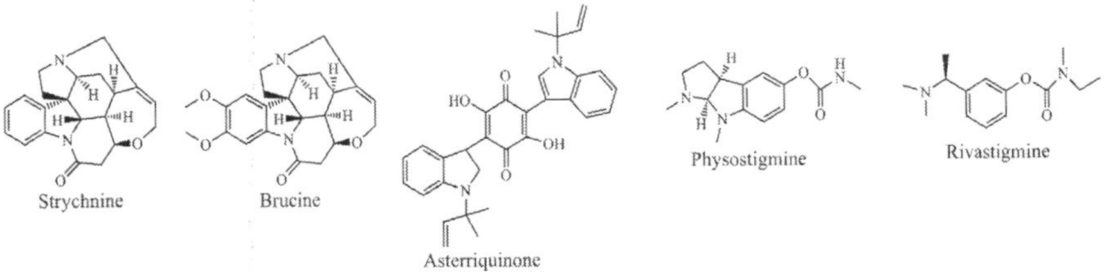

Figure 24. Indole alkaloids – strychnine, brucine, asterriquinone, physostigmine, and rivastigmine[50]

With methods of extraction developing since the 1960s, with many efforts concentrating on the efficiency and environmental impact of the process, terpenoid indole alkaloids and their pharmacological properties continue to attract attention as some of them are already marketed as anticancer, antihypertensive, or hypoglycemic agents, for example [52]. Some monoterpene indole alkaloids are in high demand, such as vinblastine (*Catharanthus roseus*) and camptothecin (*Camptotheca acuminata, Nothapodytes foetida*), currently used as chemotherapic drugs, and eyes have turned toward metabolic engineering. However, their biosynthetic pathways are still not fully elucidated and geneticians, chemists, and biotech specialists are scrambling to fill in the gaps, with resources such as cell- and organ-specific transcriptome databases on hand [53, 54].

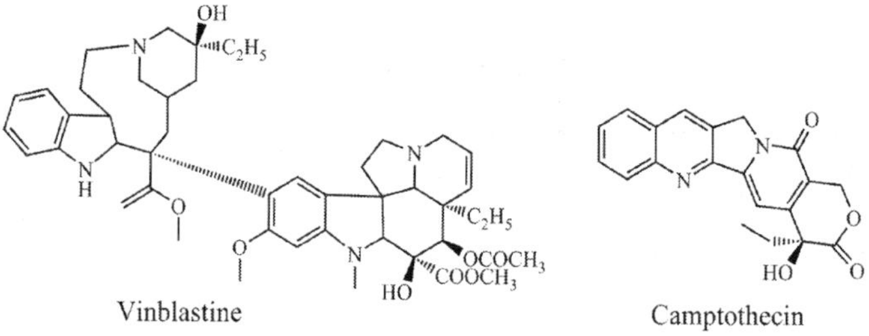

Figure 25. Indole alkaloids – vinblastine and camptothecin [50]

A novel class of indole-2-carboxylate derivatives was designed starting from the structure of pyrroloquinoline quinone, with two compounds (Figure 26) emerging as more potent antiproliferants than the reference drugs, compounds that induced PARP cleavage and increased ROS generation dose-dependence [55].

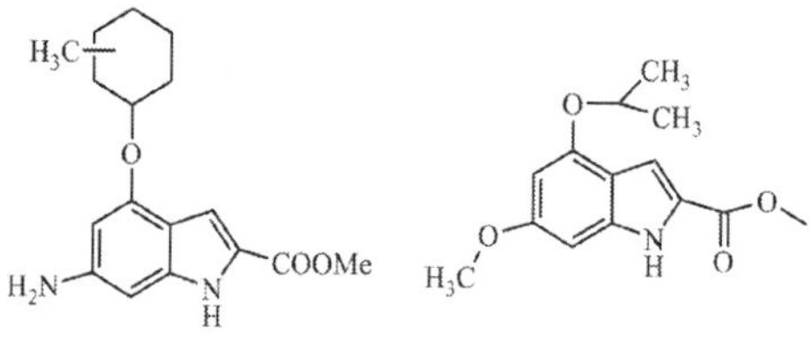

Figure 26. Indole-2-carboxylate derivatives [55]

Protein tyrosine phosphatases (PTP) are a novel, mostly untapped family of therapeutic targets, with implications in oncology (SHP2), autoimmunity (Lyp) or diabetes (PTP1B). X-ray crystal structural analysis has been performed on PTP-inhibitor complexes, revealing bicyclic benzofuran and indole-based salicylic acids as useful first steps toward the development of more potent inhibitors (Figure 27) [56].

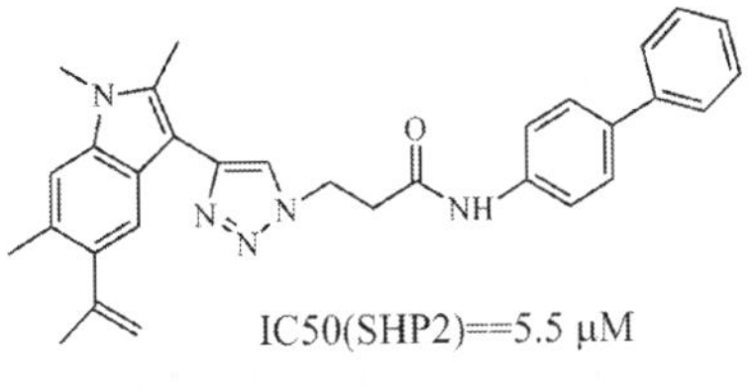

Figure 27. PTP-inhibitor [56]

Novel galantamine derivatives with indole moiety have exhibited an activity against acetyl-cholinesterase up to 95 times higher than the parent compound, with one promising lead binding in a region close to the peripheral anionic site of the enzyme, where the Ω-loop of amyloid beta peptide adheres [57].

Coronary heart disease, prevalent in industrialized regions, comes hand-in-hand with high levels of LDL-C ("bad cholesterol") and low levels of HDL-C ("good cholesterol"), treated mostly with statins, inhibitors of HMG-CoA reductase with dose-limiting hepato- and myotoxicity. However, the screening of a small indole chalcone fibrates library (Figure 28) has revealed three compounds with a more potent hypolipidemic effect than the standard drug, fenofibrate, coupled with high inhibition percentages of superoxide anions, hydroxyl radicals, and microsomal lipid-peroxidation [58].

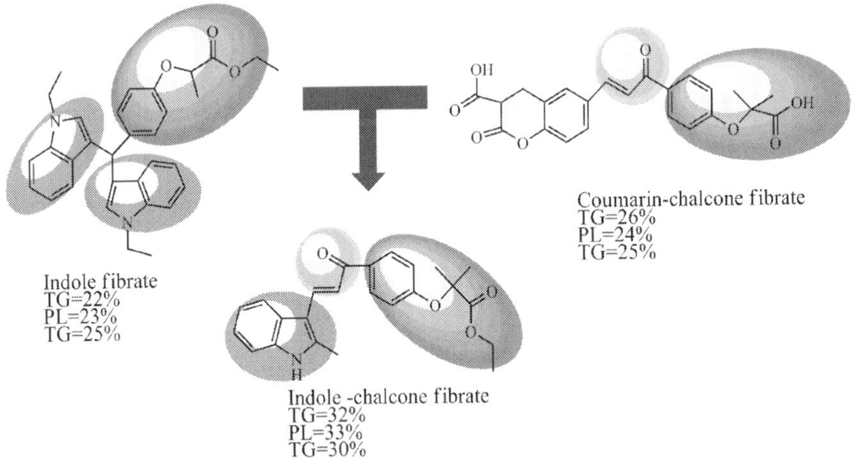

Figure 28. Indole chalcone fibrates [58]

Beneficial effects on lipid and also glucose metabolism were also reported concerning 1-(4-chlorobenzoyl)-5-hydroxy-2-methyl-3-indoleacetitic acid (GY3), which increased glucose consumption and decreased lipid accumulation through AMPK activation in hepG2 cells, with obvious implications in metabolic syndrome, type 2 diabetes, and nonalcoholic fatty liver disease [59].

Some indole derivatives also show promising antimicrobial activity. Five out of 24 bisindolylmethane Schiff base derivatives synthesized were found to specifically inhibit *Salmonella typhi, S. paratyphi A* and *S. paratyphi B,* even if the inhibition was moderate at best, when nitro or halogen substituents were introduced [43].

5.2. Natural and synthetic indolizines

Natural products derived from the indolizine core, such as castanospermine, swainsonine, or tylophorine, polyhydroxylated indolizidine alkaloids, have attracted much attention, struggles toward total synthesis, or analog design issues. Their biological properties range from the antiviral to the anticancer realm, with promising effects on autoimmune diseases [60].

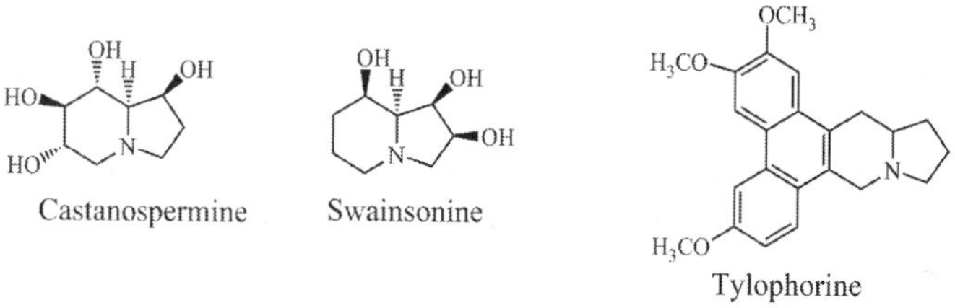

Figure 29. Indolizidine alkaloids [50]

For example, castanospermine glycoside analogs inhibit breast cancer cells MCF-7 and MDA-MB-231, inducing cell cycle arrest and apoptosis without impact on normal cell proliferation [61]. Some 5α-substituted swainsonine analogs successfully inhibit Golgi alpha-mannosidase II, a key enzyme in the N-glycosylation pathway and a potential target for cancer chemotherapy, without much loss of activity by comparison with the parent compound [62]. Tylophorine was shown to inhibit VEGFR2 tyrosine kinase activity and its downstream signaling pathways, neovascularization, tumor angiogenesis and tumor growth, molecular docking simulations indicating that it could form hydrogen bonds and have aromatic interactions within the ATP-binding region of the VEGFR2 kinase unit [63].

In the indolizine nucleus, the six-membered ring suffers from low electron density, with a subsequent charge buildup in the five-membered ring, resulting in a large dipole moment and fluorescence properties. The influence of the substituents goes a long way into predicting a blue or red-shifted fluorescence; for example, the C-2 position could carry a lot of weight [26].

The 10π conjugated planar electronic structure, exhibiting strong fluorescence properties, can be useful for DNA interaction studies. Such is the case of some indolizinylpyridinium derivatives, found to interact similarly to ethidium bromide, binding in the minor groove, but having its fluorescence partially quenched [64].

Switchable biosensors could be designed starting from Seoul-Fluor (Figure 30), an indolizine scaffold with three positions for different radicals: R1 and R2 substituents affect electronic perturbation; R3 could be a functional handle for bioconjugation, thus creating a versatile platform with tuneable emission wavelength and controllable quantum yield [65].

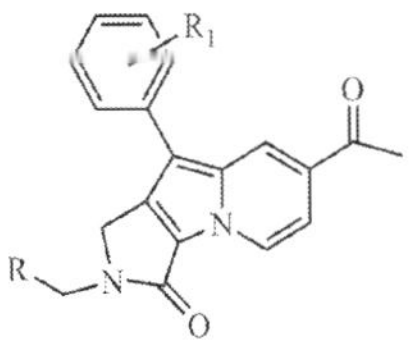

Figure 30. Indolizine biosensor: Seoul-Fluor [65]

Many indolizine derivatives have been proven to be worthy therapeutic agents, with a number of them undergoing clinical trials, notably five anticancer molecules that inhibit topoisomerase [66]. As is the case of photophysical properties, the substituents of the indolizine core can be tailored to suit the bioactivity requirements.

Following a SAR study, 49 indolizine derivatives were obtained and tested as potential HIV-1 infectivity factor inhibitors, one of which was found to exhibit an IC_{50} value of 11 μM [67].

A feature that makes indolizine derivatives attractive is the design possibilities. The facile replacement of substituents could lead to more in-depth perception toward their effect upon desired bioactivity, solubility, or other properties sought [66].

Two new classes of indolizines fused with phenantroline skeletons were designed and synthesized, obtaining compounds with a coplanar structure, potentially able to interact with DNA through an intercalation mechanism, compounds that also possess good solubility in microbiological medium. Furthermore, one of the compounds exhibit, under aerobic conditions, activity against *M. tuberculosis* H37Rv, with an IC_{50} = 67 μM. Two other compounds had a selective and significant antiproliferative activity (around 50%) against two breast cancer cell lines (MCF7 and T-47D) (Figure 31) [68].

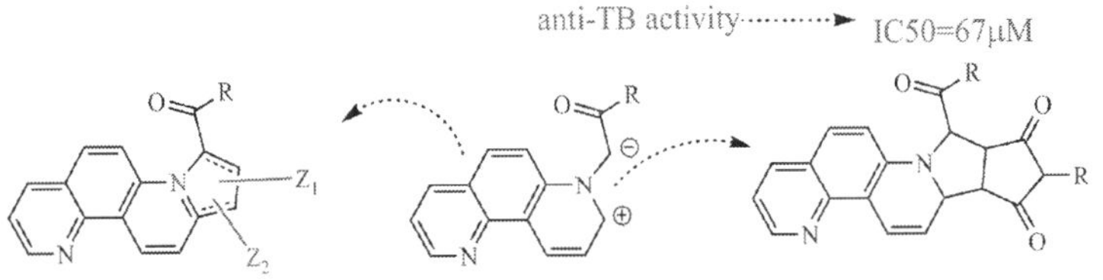

Figure 31. Indolizines fused with phenantroline skeletons active against M. tuberculosis and breast cancer [68]

During rational design efforts, concentrated on the identification of potential farnesyltransferase inhibitors (implications with respect to oncogenic Ras proteins), the replacement of the triazole unit with the indolizine nucleus resulted in IC_{50}s in the low micromolar range. The substituents' influence on bioactivity and pharmacokinetic parameters was also investigated [69].

Rational design could be pushed even further, employing 3D-QSAR studies to yield pharmacophore models, as is the case for 15-lipoxygenase inhibitory activity. For this purpose, 47 indolizines with anti-15-LOX activity were used to obtain a statistically significant model [70].

The similarity of the two heterocycles has motivated researchers since 1967, when Harrell and Doerge postulated that indolizine analogs of bioactive indoles could possess similar or improved potency [66]. Such an endeavor was attempted with the synthesis of 1-(2-aminoethyl)-3-benzyl-7-methoxy-2-methylindolizine, an analog of indole derivative benanserin, the replacement of indole with indolizine proving to have no effect on anti-acetylcholinesterase activity but diminished the antihistamine and anti-5-hydroxytryptamine activity [66].

Ramatroban, 3-((3R)-3-{[(4-fluorophenyl)sulfonyl]amino}-1,2,3,4-tetrahydro-9H-carbazol-9-yl)propanoic acid, is a well-known prostaglandin D2 inhibitor and thromboxane receptor antagonist. Researchers from Merck and Amira have presented numerous ramatroban analogs with the indole moiety replaced with indolizine and aza-indole, both proving potential during SAR studies. Among them, a 4-aza-indole derivative (Figure 32) inhibited hCRTH2 with an IC of 6 nM and was active in a murine OVA-induced lung inflammation model [71-74].

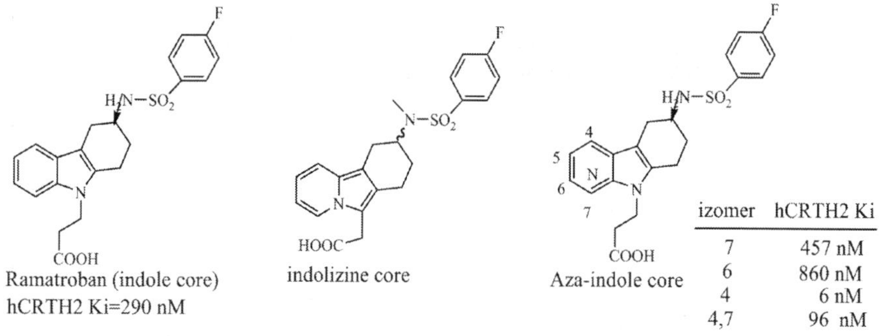

izomer	hCRTH2 Ki
7	457 nM
6	860 nM
4	6 nM
4,7	96 nM

Figure 32. Tricyclic CRTH2 antagonist [74]

LE 300 is a selective antagonist for dopamine D1/D5 and serotonin 5-HT (2A) receptors, bearing an azecine ring fused to an indole ring on one side and a benzene moiety on the other side. New analogs of this compound were prepared, namely pyrrolo[2,3-g]indolizine, pyrrolo[3,2-a]quinolizine rings and their corresponding dimethylpyrrolo[2,3-d]azonine, and dimethylpyrrolo[2,3-d]azecine [75]. The study concludes that the indolizine and quinolizine derivatives show no activity concerning the receptors analyzed, while their azonine and azecine counterparts exhibited only weak antagonistic effects for serotonin and histamine receptors, remaining nonresponsive toward the four dopamine receptors tested.

Secretory phospholipases A2 (sPLA2s) is successfully inhibited by substituted indole and indolizine derivatives, as previously reported by Lilly and Shinogi researchers, with compounds like indoxam and Me-indoxam exhibiting favorable pharmacokinetic profiles [76-79]. Oslund et al. have prepared a set of benzo-fused analogs, among which they identified a compound that was the first reported potent inhibitor of groups IID and IIF sPLA2s and the most generally potent sPLA2 inhibitor reported to date (Figure 33) [80].

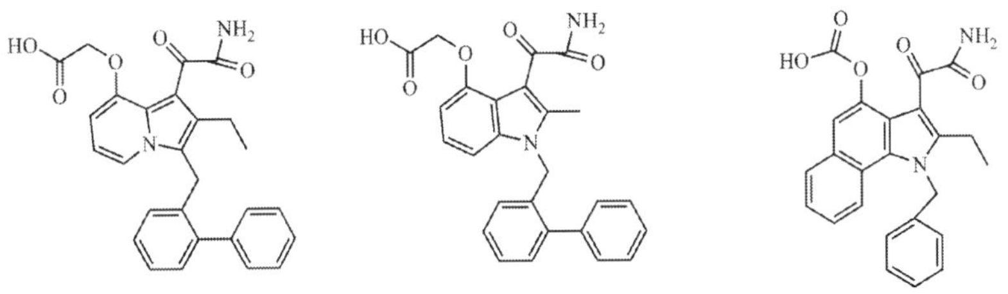

Indoxam Me-indoxam

Figure 33. Substituted indole, indolizine, and benzo-fused indole inhibitors against human and mouse sPLA2 [80]

James et al. have prepared a series of indole and indolizine-glyoxylamides (Figure 34) and subsequently tested the compounds' cytotoxicity against cancer cell lines, identifying high antiproliferative activities, even in the case of multidrug-resistant phenotypes. After searching among numerous cores, with the goal to replace indole, the researchers synthesized a novel class of cancer agents with an indolizine core, with a lead compound that proved effective against multidrug-resistant cell lines such as MES-SA/DX5 and HL60/TX1000, resistant to treatment with Taxol [81].

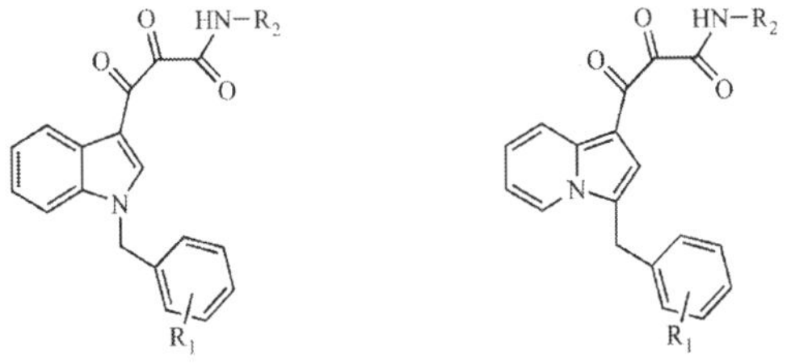

Figure 34. Indole- and Indolizine-glyoxylamide derivatives [81]

6. Conclusions

Considering the growing interest for biologically active compounds, we believe that in the future the search for novel indole and indolizine derivatives will result in the emergence of new synthesis pathways and new and unexplored biologically active derivatives with pyrrole moieties.

Taking into account the importance of anticancer drugs, like vinblastine, irinotecan, topotecan, or camptothecin, the development of biologically active derivatives of new natural lead compounds containing indole and indolizine nucleus might be helpful in the design and development of novel and more potent anticancer drugs, antiviral agents, intercalating agents.

With the remarkable number of approved indole-containing drugs as well as the importance of the indolizine moiety, it can be easily concluded that indole and indolizine derivatives offer perspectives on how pyrrole scaffolds might be exploited in the future as bioactive molecules against a broad range of diseases.

Acknowledgements

This work was supported by a grant of the Romanian National Authority for Scientific Research, CNCS – UEFISCDI, project number PN-II-ID-PCE-2011-3-0226

Author details

Ioana Otilia Ghinea and Rodica Mihaela Dinica*

*Address all correspondence to: rodinica@ugal.ro

"Dunărea de Jos" University of Galati, Department of Chemistry, Physics and Environment, Galati, Romania

References

[1] Yurovskaya MA, Alekseyev RS. New perspectives on classical heterocyclic rections involving pyrrole derivatives (Review). *Chem. Het. Comp.* 2014;49(10):1400-1425.

[2] Niu T, Huang L, Wu T and Zhang Y. FeCl3-promoted alkylation of indoles by enamides. *Org. Biomol. Chem.* 2011;9:273.

[3] Ganguly NC, Mondal P and Barik SK. Iodine in aqueous micellar environment: a mild effective ecofriendly catalytic system for expedient synthesis of bis(indolyl)methanes and 3-substituted indolyl ketones. *Green Chem. Lett. Rev.* 2012;5(1):73-81.

[4] Ieronimo G, Mondelli A, Tibiletti F, Maspero A, Palmisano G, Galli S et al. A simple, efficient, regioselective and one-pot preparation of N-hydroxy- and N-O-protected

hydroxyindoles via cycloaddition of nitrosoarenes with alkynes. Synthetic scope, applications and novel by-products. *Tetrahedron*. 2013;69:10906-10920.

[5] Tabera DF, Tirunaharib PK. Indole synthesis: a review and proposed classification. *Tetrahedron*. 2011;67(38):7195-7210.

[6] Matczak R, Koszarna B, Gryko DT. 1-(imidazo-1-yl)-indolizines, two-step synthesis and optical properties. *Tetrahedron*. 2014;70:7006-7009.

[7] Caira MR, Popa MM, Draghici C, Barbu L, Dumitrescu D, Dumitrascu F. 7,8,9,10-tetrahydropyrrolo[2,1-a]isoquinolines in the search for new indolizine derivatives. *Tet. Lett*. 2014;55:5635-5638.

[8] Sharma V, Kumar V. One-pot synthesis of animoindolizines and chalcones using CuI/CSP nanocomposites with anomalous selectivity under green conditions. *Med. Chem. Res*. 2014;23:3593-3606.

[9] Chernyak N, Tilly D, Li Z and Gevorgyan V. Cascade carbopalladation-annulation approach toward polyciclic derivatives of indole and indolizine. *ARKIVOC*. 2011;76-91.

[10] Singh MS, Chowdhury S. Recent developments in solvent-free multicomponent reactions: a perfect synergy for eco-compatible organic synthesis. *RSC Adv*. 2012;2:454-4592.

[11] Outlaw VK, d`Andrea FB and Townsend CA. One-pot synthesis of highly-substituted N-fused heteroaromatic bicycles from azole aldehydes. *Org. Lett*. 2015;17(8): 1822-1825.

[12] Liu H, Chen H, Li Y, Lv Y, Cai J, Ji M. A novel and one-pot method for synthesis of unprecedented 3,3-dimethyl-2-amide indoles under metal-free conditions. *Tet. Lett*. 2015;56:2332-2335.

[13] Song Y-L, Wu F, Zhang C-C, Liang G-C, Zhou G, Yu J-J. Ionic liquid catalyzed synthesis of 2-(indole-3-yl)-thiochroman-4-ones and their novel antifungal activities. *Bioorg. Med. Chem. Lett*. 2015;25:259-261.

[14] Ekbote SS, Deshmukh KM, Qureshi ZS and Bhanage BM. Polyvinylsulfonic acid as a novel Bronsted acid catalyst for the synthesis of bis(indolyl)metahnes. *Green Chem. Lett. Rev*. 2011;4(2):177-183.

[15] Surasania R, Kalitaa D and Chadrasekhar KB. Indion Ina 225H resin as a novel, selective, recyclable, eco-benign heterogenous catalyst for the sythesis of bis(indolyl)methanes. *Green Chem. Lett. Rev*. 2013;6(2):113-122.

[16] Singh R, Panda G. An overview of synthetic approaches for heterocyclic steroids. *Tetrahedron*. 2013;69:2853-2884.

[17] Sandeep C, Padmashali B, Kulkarni RS. Efficient synthesis of indolizines and new imidazo[1,2-a]pyridines via the expected cyclization of aromatic cycloimmonium

ylides with electron deficient alkynes and ethyl cyanoformate. *Tet. Lett.* 2013;54:6411-6414.

[18] Dontsova NE, Nesterov VN, Shestopalov AM. Effect of solvent nature on the regioselectivity of the reactions of pyridinium ylides with E-1,2-di(alkylsulfonyl)-1,2-dichloroethene. From the reaction of 1,3-dipolar cycloaddition to the reaction of nucleophilic addition-elimination (AdN-E1,5). *Tetrahedron.* 2013;69:5016-5021.

[19] Mishra S, Naskar B, Ghosh R. CuCl catalyzed green and efficient one-pot synthesis of aminoindolizine frameworks via three-component reactions of aldehydes, secondary amines, and terminal alkynes in PEG. *Tet. Lett.* 2012;53:5483-5487.

[20] Li L, Chua WKS. One-pot multistep synthesis of 3-aminoindolizine derivatives. *Tet. Lett.* 2011;52:1392-1394.

[21] Ziyaadini M, Hazeri N, Maghsoodlou MT, Habibi-Khorassani SM, Willis AC. A novel one-pot domino reaction for the synthesis of 2-acetyl-3(phenylamino)indolizine-1-carboxamide derivatives. *Tet. Lett.* 2011;52:5774-5776.

[22] Mao Z, Li X, Lin X, Lu P, Wang Y. One-pot multicomponent synthesis of polysubstituted indolizines. *Tetrahedron.* 2012;68:85-91.

[23] Kucukdisli M and Opatz T. One-pot synthesis of polysubstituted indolizines by an addition/cycloaromatization sequence. *J. Org. Chem.* 2013;78:6670-6676.

[24] Lee JH, Kim I. Cycloaromatization approach to polysubstituted indolizines from 2-acetylpyrroles: decoration of the pyridine unit. *J. Org. Chem.* 2013;78:1283.

[25] Park S, Kim I. Electron-withdrawing group effect in aryl gropu of allyl bromides for the successful synthesis of indolizines via a novel [3+3] annulation approach. *Tetrahedron.* 2015;71:1982-1991.

[26] Ge YQ, Gong XY, Song GJ, Cao XQ, Wang JW. The synthesis, characterization and optical properties of novel 2-acyl 6-arylindolizines. *Spectrochimica Acta A: Mol. Biomol. Spectrosc.* 2015;135:7-13.

[27] Pandya AN, Fletcher JT, Villa EM, Agrawal DK. Silver-mediated synthesis of indolizines via oxidative C-H functionalization and 5-endo-dig cyclization. *Tet. Lett.* 2014;55:6922-6924.

[28] Yang Y, Gao M, Zhang D-X, Wu L-M, Shu W-M, Wu A-X. Convergent domino synthesis of 1,2,3-triaroylindolizines from methyl ketones and pyridines via self-division of labor strategy. *Tetrahedron.* 2012;68:7338-7344.

[29] Hazra A, Mondal NB, Maity A, Naskar S, Saha P, Paira R et al. Amberlite IRA-402 (OH) ion exchange resin mediated synthesis of indolizines, pyrrolo [1,2-a] quinolines and isoquinolines: antibacterial and antifungal evaluation of the products. *Eur. J. Med. Chem.* 2011;46:2132-2140.

[30] Gopal MS and Anitha I. Aqueous phase microwave synthesis of some bisindolizines. *Int. J. Chem. Std.* 2014;2(4):35-38.

[31] Dinica RM, Furdui B, Ghinea IO, Bahrim G, Bonte S and Demeunynck M. Novel one-pot green synthesis of indolizines biocatalysed by Candida antarctica lipases. *Mar. Drugs.* 2013;11:431-439.

[32] Nelson AC, Kalinowski ES, Jacobson TL, Grundt P. Formation of tryptanthrin compounds upon Oxone-induced dimerization of indole-3-carbaldehydes. *Tet. Lett.* 2013;54:6804-6806.

[33] Kumar PP, Reddy YD, Reddy CVR, Devi BR and Dubey PK. Green synthesis of novel (E)-2-(1,4-dioxo-1,2,3,4-tetrahydrophthalazine-2-carbonyl)-3-(1H-indol-3-yl)acrylonitriles. *Green Chem. Lett. Rev.* 2014;7(4):322-329.

[34] Meshram HM, Thakur PB and Bejjam MB. An efficient synthesis of hemiaminal of indoles by using tetrabutylammonium fluoride (TBAF) in water as a reusable reaction media. *Green Chem. Lett. Rev.* 2013;6(1):95-100.

[35] Hoseini-Sarvari M and Parhizgar G. Regioselective Friedel-Crafts alkylation of indoles with epoxides using nano MgO. *Green Chem. Lett. Rev.* 2012;5(3):439-449.

[36] Talukdar D and Thakur AJ. A green synthesis of symmetrical bis(indol-3-yl)methanes using phosphate-impregnated titania catalyst under solvent free grinding conditions. *Green Chem. Lett. Rev.* 2013;6(1):55-61.

[37] Kamalraja J, Perumal PT. Microwave assisted InCl3 mediated regioselective synthesis of highly functionalized indolylpyran under solvent-free condition and its chemical transformation to indolyltriazolylpyran hybrids. *Tet. Lett.* 2014;55:3561-3564.

[38] Siddiqui ZN, Farooq F and Musthafa TNM. A highly efficient, simple, and ecofriendly microwave induced synthesis of indolyl chalcones and pyrazolines. *Green Chem. Lett.* Rev. 2011;4(1):63-68.

[39] Vieira BM, Thurow S, Brito JS, Perin G, Alves D, Jacob RG et al. Sonochemistry: an efficient alternative to the synthesis of 3-selanylindoles using CuI as catalyst. *Ultrasonics Sonochem.* 2015;27:192-199.

[40] Heravia MM, Farda MV and Faghihi Z. Heteropoly acids-catalyzed organic reactions in water: doubly green reactions. *Green Chem. Lett. Rev.* 2013;6(4):282-300.

[41] Proenc F, Costa M. A one-pot synthesis of substituted pyrido[2,3-b]indolizines. *Tetrahedron.* 2011;67:1071-1075.

[42] Zhanga M-Z, Chena Q, Yang G-F. A review on recent developments of indole-containing antiviral agents. *Eur. J. Med. Chem.* 2015;89:421-441.

[43] Imran S, Taha M, Ismail NH, Khan KM, Naz F, Hussain M and Tauseef S. Synthesis of novel bisindolylmethane Schiff bases and their antibacterial activity. *Molecules.* 2014;19:11722-11740.

[44] Khan N, Mukhtar H. Dietary agents for prevention and treatment of lung cancer. *Cancer Lett.* 2015;359:155-164.

[45] Riso P, del Bo C. Preventive effects of broccoli bioactives: role on oxidative stress and cancer risk. In: Preedy VR, editor. *Cancer: Oxidative Stress and Dietary Antioxidants.* Elsevier Inc; 2014. p. 115-125.

[46] Yu TY, Pang WJ, Yang G-S. 3,3`-Diindolylmethane increases bone mass by suppressing osteoclastic bone resorption in mice. *J. Pharmacol. Sci.* 2015;127:75-82.

[47] Deng W, Zong J, Wei L, Guo H, Cheng Z, Zhang R, et al. 3,3′-Diindolylmethane improves myocardial energy metabolism imbalance induced by pressure overload via AMPKα in mice. *Int. J. Cardiol.* 2014;177:235-237.

[48] Saraswati S, Agrawal SS. Brucine, an indole alkaloid from Strychnos nux-vomica attenuates VEGF-induced angiogenesis via inhibiting VEGFR2 signaling pathway in vitro and in vivo. *Cancer Lett.* 2013;332(1):83-93.

[49] Carbone A, Parrino B, Barraja P, Spanò V, Cirrincione G, Diana P, et al. Synthesis and antiproliferative activity of 2,5-bis(3′-Indolyl)pyrroles, analogues of the marine alkaloid Nortopsentin. *Mar. Drugs* 2013;11:643-654.

[50] Aniszewski T. Biology of alkaloids. In: *Alkaloids – Chemistry, Biology, Ecology and Applications.* 2nd ed. Elsevier B.V.; 2015. p. 196-257.

[51] Bhadra S, Dalai MK, Chanda J, Mukherjee PK. Evaluation of bioactive compounds as acetylcholinesterase inhibitors from medicinal plants. In: *Evidence-Based Validation of Herbal Medicine.* Elsevier Inc.; 2015. p. 274-300.

[52] Goldhaber-Pasillas GD, Choi YH Verpoorte R. New methods of analysis and investigation of terpenoid indole alkaloids. In: *Advances in Botanical Research – New Light on Alkaloid Biosynthesis and Future Prospect.* 2013. p. 233-272.

[53] Sirikantaramas S, Yamazaki M, Saito K. Camptothecin: Biosynthesis, biotechnological production and resistance mechanism(s). In: *Advances in Botanical Research – New Light on Alkaloid Biosynthesis and Future Prospect.* 2013. p. 139-161.

[54] Salim V, De Luca V. Towards complete elucidation of monoterpene indole alkaloid biosynthesis pathway: Catharanthus roseus as a pioneer system. In: *Advances in Botanical Research – New Light on Alkaloid Biosynthesis and Future Prospect.* 2013. p. 1-37.

[55] Ji X-Y, Xue S-T, Zhan Y-C, Shen J-J, Wu L-T, Jin J, et al. Design, synthesis and antiproliferative activity of a novel class of indole-2-carboxylate derivatives. *Eur. J. Med. Chem.* 2014;83:409-418.

[56] He Y, Zeng L-F, Yu Z-H, He R, Liu S, Zhang Z-Y. Bicyclic benzofuran and indole-based salicylic acids as protein tyrosine phosphatase inhibitors. *Bioorg. Med. Chem.* 2012;20:1940-1946.

[57] Atanasova M, Stavrakov G, Philipova I, Zheleva D, Yordanov N, Doytchinova I. Galantamine derivatives with indole moiety: Docking, design, synthesis and acetylcholinesterase inhibitory activity. *Bioorg. Med. Chem.* 2015;23:5382-5389.

[58] Sashidhara KV, Dodda RP, Sonkar R, Palnati GR, Bhatia G. Design and synthesis of novel indole-chalcone fibrates as lipid lowering agents. *Eur. J. Med. Chem.* 2014;81:499-509.

[59] Si M, Yan Y, Tang L, Wu H, Yang B, He Q, Wu H. A novel indole derivative compound GY3 improves glucose and lipid metabolism via activation of AMP-activated protein kinase pathway. *Eur J Pharmacol*. 2013;698:480-488.

[60] Koskinen AMP and Kallatsa OA. Polyhydroxylated indolizidine alkaloids – an efficient synthesis of 1-deoxy-8,8a-di-epi-castanospermine. *Tetrahedron* 2003;59:6947-6954.

[61] Allan G, Ouadid-Ahidouch H, Sanchez-Fernandez EM, Risquez-Cuadro R, Fernandez JMG, Ortiz-Mellet C, Ahidouch A. New castanospermine glycoside analogues inhibit breast cancer cell proliferation and induce apoptosis without affecting normal cells. *PLoS ONE*. 2013;8(10):76411.

[62] Kuntz DA, Nakayama S, Shea K, Hori H, Uto Y, Nagasawa H, Rose DR. Structural investigation of the binding of 5-substituted swainsonine analogues to Golgi alpha-mannosidase II. *Chembiochem*. 2010;11(5):673-680.

[63] Saraswati S, Kanaujia PK, Kumar S, Kumar R and Alhaider AA. Tylophorine, a phenanthraindolizidine alkaloid isolated from Tylophora indica exerts antiangiogenic and antitumor activity by targeting vascular endothelial growth factor receptor 2–mediated angiogenesis. *Mol Cancer*. 2013;12:82.

[64] Gradinaru R-V, Luca A-M, Cretescu I, Danac R. Fluorescent conjugates: pH stability, dye-DNA interaction and biological activity. *REV. CHIM.* (Bucharest). 2010;61:9.

[65] Kim E, Lee Y, Lee S and Park SB. Discovery, understanding, and bioapplication of organic fluorophore: A case study with an indolizine-based novel fluorophore, Seoul-Fluor. *Acc. Chem. Res.* 2015;48:538-547.

[66] Sharma V, Kumar V. Indolizine: a biologically active moiety. *Med Chem Res.* 2014;23:3593-3606.

[67] Huang W, Zuo T, Luo X, Jin H, Liu Z, Yang Z, et al. Indolizine derivatives as HIV-1 VIF–elonginC interaction inhibitors. *Chem. Biol. Drug. Des.* 2013;81:730-741.

[68] Danac R, Al Matarneh CM, Shova S, Daniloaia T, Balan M, Mangalagiu II. New indolizines with phenanthroline skeleton: Synthesis, structure, antimycobacterial and anticancer evaluation. *Bioorg. Med. Chem.* 2015;23:2318-2327.

[69] Dumea C, Belei D, Ghinet A, Dubois J, Farce A, Bicu E. Novel indolizine derivatives with unprecedented inhibitory activity on human farnesyltransferase. *Bioorg. Med. Chem. Lett.* 2014;24:5777-5781.

[70] Sharma V, Kumar V. Pharmacophore mapping studies on indolizine derivatives as 15-LOX inhibitors. Bulletin of Faculty of Pharmacy, Cairo University. 2015;53:63-68.

[71] Stearns BA, Baccei C, Bain G, Broadhead A, Clark RC, Coate H, et al. Novel tricyclic antagonists of the prostaglandin D2 receptor DP2 with efficacy in a murine model of allergic rhinitis. *Bioorg.Med.Chem Lett.* 2009;19:4647.

[72] Gervais FG, Sawyer N, Stocco R, Hamel M, Krawczyk C, Sillaots S, et al. Pharmacological characterization of MK-7246, a potent and selective CRTH2 (chemoattractant receptor-homologous molecule expressed on T-Helper Type 2 Cells) Antagonist. *Mol. Pharmacol.* 2011;79:69-76.

[73] Gallant M, Beaulieu C, Berthelette C, Colucci J, Crackower MA, Dalton C, et al. Discovery of MK-7246, a selective CRTH2 antagonist for the treatment of respiratory diseases. *Bioorg. Med. Chem. Lett.* 2011;21:288.

[74] Tumey N. CRTH2 antagonists. In: Levin JI, Laufer S, editors. *Anti-Inflammatory Drug Discovery*. RSC Publishing; 2012. p. 109-112.

[75] Rostom SA. Novel fused pyrrole heterocyclic ring systems as structure analogs of LE 300: Synthesis and pharmacological evaluation as serotonin 5-HT(2A), dopamine and histamine H(1) receptor ligands. *Achiv Der Pharmazie.* 2010;343:73-80.

[76] Dillard RD, Bach NJ, Draheim SE, Berry DR, Carlson DG, Chirgadze NY, et al. Indole inhibitors of human nonpancreatic secretory phospholipase A2. 1. Indole-3-acetamides. *J. Med. Chem.* 1996;39:5119-5136.

[77] Dillard RD, Bach NJ, Draheim SE, Berry DR, Carlson DG, Chirgadze NY, et al. Indole inhibitors of human nonpancreatic secretory phospholipase A2. 2. Indole-3-acetamides with additional functionality. *J. Med. Chem.* 1996;39:5137-5158.

[78] Draheim SE, Bach NJ, Dillard RD, Berry DR, Carlson DG, Chirgadze NY, et al. Indole inhibitors of human nonpancreatic secretory phospholipase A2. 3. Indole-3-glyoxamides. *J. Med. Chem.* 1996;39:5159-5175.

[79] Hagishita S, Yamada M, Shirahase K, Okada T, Murakami Y, Ito Y, et al. Potent inhibitors of secretory phospholipase A2: Synthesis and inhibitory activities of indolizine and indene derivatives. *J. Med. Chem.* 1996;39:3636-3658.

[80] Oslund RC, Cermak N and Gelb MH. Highly specific and broadly potent inhibitors of mammalian secreted phospholipases A2. *J. Med. Chem.* 2008;51(15):4708-4714.

[81] James DA, Koya K, Li H, Liang G, Xia Z, Ying W, et al. Indole- and indolizine-glyoxylamides displaying cytotoxicity against multidrug resistant cancer cell lines. *Bioorg. Med. Chem. Lett.* 2008;18(6):1784-1787.

Chapter 6

Synthesis of Nitriles – Synthesis of 4-Cyano Pyrazole, 5-Aminopyrazole Derivatives and the Deamination of 5-Aminopyrazole Derivatives

Raghunath Toche

Additional information is available at the end of the chapter

http://dx.doi.org/10.5772/64050

Abstract

Chemoselective reaction on 3-dimethylamino-2-aroyl-propenenitrile and hydrazine in acidic medium yields 4-cyano pyrazole, where as in basic medium yields 5-amino pyrazoles as major product.

Keywords: 4-Cyanopyrazole, 5-aminopyrazole, Deamination, Isopentyl nitrite, Chemoselective reactions

1. Introduction

Pyrazole is an organic compound having a molecular formula $C_3H_4N_2$, pentatomic heterocycle with a nitrogen heteroatom, having a five member ring structure with three carbon and adjacent two nitrogen atoms. Pyrazoles rarely occur in nature; in 1959, β-(1-pyrazolyl) alanine was isolated from the seeds of water melons (*Citurllus lanatus*) (L. Fowden). Pyrazoles exhibit wild range of biological activities such as anti-diabetic, antiviral, anti-cancer, anti-inflammatory, antibacterial, and antifungal activities).

History: Ludwig Knorr (1883) has given the name pyrazole to this class of compounds. The reduced forms of pyrazoles are pyrazoline and pyrazolidine. The substituted derivatives of pyrazole has been used in medicines and in other technical applications.

1.1. Physical properties

Pyrazole is a colorless solid, boiling points (b.p), 186-188°C, melting point (m.p.), 67-70 °C, a weak base Pk_b = 11.5 (pK_a of the conjugated acid 2.49 at 25 °C, Mol. Wt. 68.0776 g/mol, and soluble in water

1.2. Chemistry of pyrazole

The high m.p. and b.p. of pyrazole compared with 1-alkyl or aryl substituted pyrazoles are due to intermolecular hydrogen bonding which results in the dimmer. It is a tautomeric substance. Pyrazole is a weak basic and forms salts with inorganic acids; the imino hydrogen may be replaced by an acyl group.

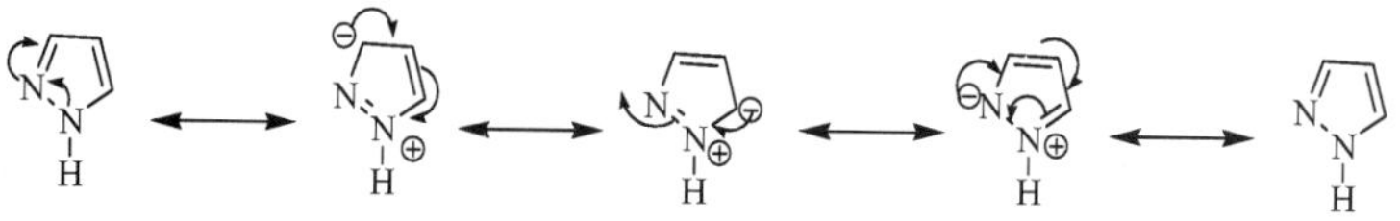

Figure 1. Resonating structures for pyrazole

Pyrazole resistant to oxidation and reduction reaction due to loss of aromaticity, but may be hydrogenated catalytically, first to pyrazoline, and then to pyrazolidine. Both of these compounds are stronger bases than pyrazole.

Oxidation: Pyrazole ring system is resistant to oxidizing agents but the side chain may be oxidized to carboxylic acid group in the presence of potassium permanganate.

Reduction: Pyrazole ring system can be reduced with molecular hydrogen and metal catalyst to pyrazole and pyrazolidine both are stronger bases than pyrazole.

1.3. Alkylation and acylation

The free N-H group in pyrazole can be alkylated with alkylating agents such as alkyl halides, diazomethane, and dimethyl sulfate or acylated using acid chloride and acetic anhydride.

Electrophilic aromatic substitutions: Pyrazole is an aromatic compound that exhibits all the properties of aromatic compounds such as electrophilic substitution reactions e.g. halogenation, nitration, sulfonation, etc., in neutral or in basic medium, but not in acidic medium. The substitution occurs at C_4-position through the formation of arenium ion intermediate.

Reactions of pyrazoles with nucleophilies: The presence of a strong electron-withdrawing group on pyazole assists nucleophilic substitution.

General synthesis:-

1. Pyrazoles and their derivatives were synthesized from hydrazine or its derivatives and a 1,3-dicarbonyl compound using an acid catalyst, the reaction is also known as Knorr pyrazole synthesis.

2. Sucrow reported the synthesis of pyrazole using monomethyl hydrazones of dialkyloxalacetates.

3. Hart and Brew Baker have described the cyclization of 1,3- bis(diazopropane) to pyrazole by a concerted intermolecular 1,3-dipolar cycloaddition reaction.

http://dx.doi.org/10.5772/64050

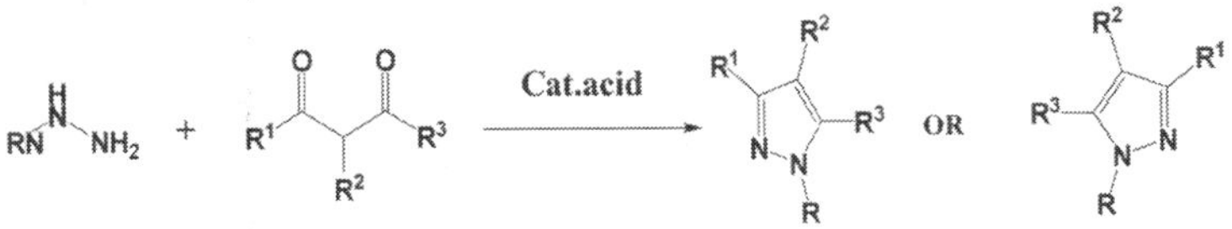

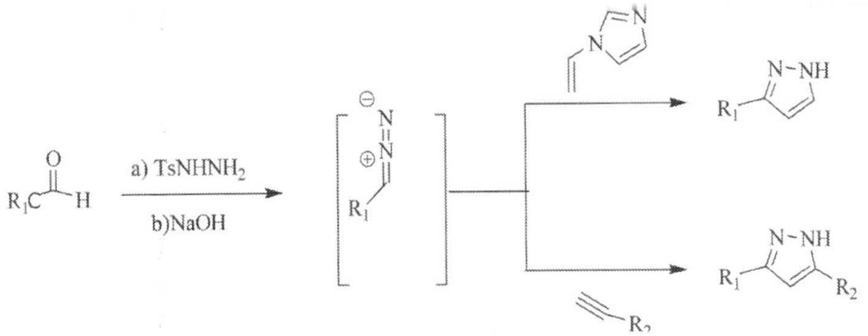

4. Pyrazoles are prepared by the action of hydrazine on 1,3-di-functional derivatives, such as carbonyl group, which can be replaced by a three-member ring, usually oxirane-aziridine-β-substituted-pyrrole-indole derivatives.

5. The addition of diazo compound to acetylenes gives pyrazole derivatives.

The same reaction as applied to olefin leads to dihydropyrazoles which are termed pyrazolines.

6. Reaction of hydrazine and their derivatives with α, β-unsaturated aldehyde / ketones yields pyrazolines.

1.4. Pharmacological interest

Pyrazole nucleus constitutes a number of sub-structures of natural products and biologically active compounds. Several derivatives of these systems find use in medicine described as follows:

Derivatives of pirolidine as drugs: Piracetam (Nootropilum) polyvinylpyrrolidone used for dementia and cognitive problems such as a chronic or persistent disorder of the mental processes caused by brain disease or injury and marked by memory disorders, personality changes, and impaired reasoning.

Derivatives of pyrazolone-5 as drugs: Phenazone (antipyrine) *Antipyrine* and benzocaine otic are *used* to relieve ear pain and swelling caused by middle ear infections. The dipyrone (metamizole sodium) is an organic *sodium* salt of antipyrine substituted at C-4 by a methyl(sulfonatomethyl)amino group, commonly *used* as a powerful analgesic and antipyretic. The budirol (propiphenazonum) is an analgesic efficacy.

Derivatives of pyrazolidine-3, 5-dione as drugs: Phenylbutazone, tribuzonum, kebuzone. Derivatives of pirolidine as drugs: Piracetam (Nootropilum), polyvinylpyrrolidone used for dementia cognitive problems such as a chronic or persistent disorder of the mental processes caused by brain disease or injury and marked by memory disorders, personality changes, and impaired reasoning.

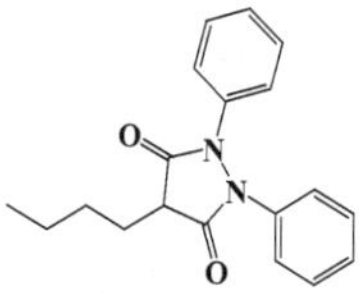	**Phenylbutazone** is 4-butyl-1, 2-diphenyl pyrazolidine 3,5-dione used as analgesic, antiinflammatory, and antipyretic drugs, and also used for the treatment of rheumatic disorder.
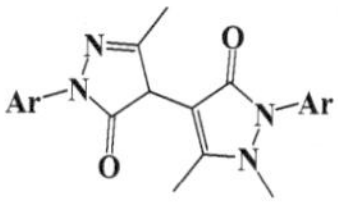	**Forbisen** is 2, 2', 3, 3'-tetramethyl-1, 1'-diphenyl-4, 4'-bi-3, 3'-pyrazoline-5, 5'-dione a by-product obtained in the manufacture of antipyrine, and has been used in bovine anaplasmosis.
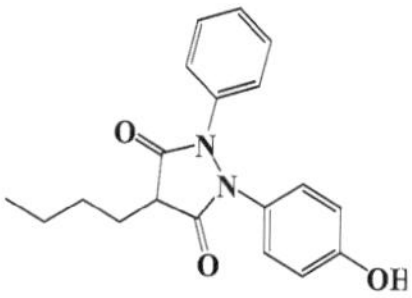	**Oxyphenbutazone** is used for the treatment of inflammation of the eyes and also is one of the active metabolite of phenylbutazone.
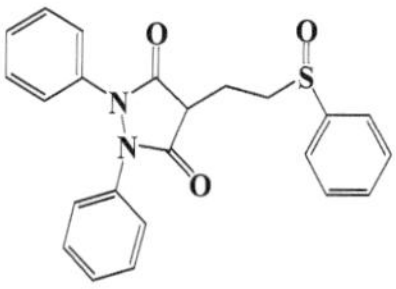	**Sulphinpyrazone** is an analogue of phenylbutazone having 2-phenylsulphinylethyl group in the place of *n*-butyl group at C_4-position; promotes excretion of uric acid and urate by inhibiting their tubular reabsorption.
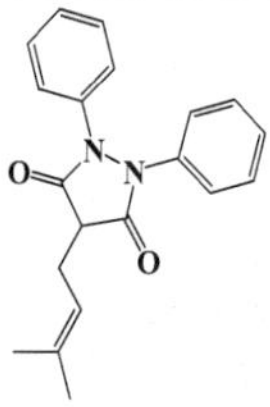	**Feprazone**: Structurally, it is similar to phenylbutazone except that the former is having a 3 methylbutenyl substituent at C_4-position of pyrazoline-2,5-dione skeleton in place of a butyl substituent. Feprazone also finds use in the treatment of rheumatic disorders.
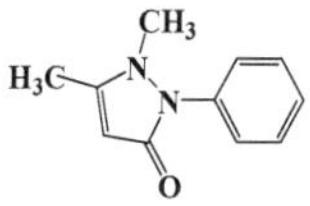	**Phenazone** is a pyrazoline derivative, chemically 2,3-dimethyl-1-phenyl 3-pyrazolin-5-one, available in white crystals or white crystalline powder soluble in water. Phenazone is well known for its analgesic and antipyretic actions.
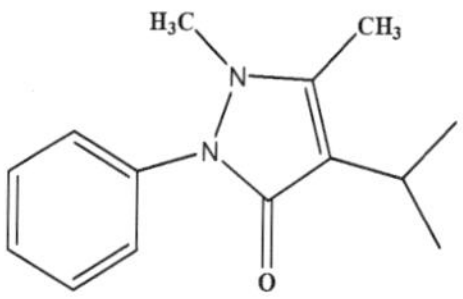	**Propylphenazone** (4-isopropyl 2,3-dimethyl-1-phenyl-3-pyrazolin-5-one) is phenazone derivative with C_4-isopropyl side chain having analgesic properties.

5-Pyrazolone derivatives are also used as cotton azo dye to improve quality such as brightness and light fastness property.

The 5-aminopyrazole system represents an important hetero-cyclic compound having considerable interest due their long history of applications in the pharmaceutical and agro-chemical industries [1-4].

Literature reports over the past hundred years and their chemistry have been reviewed in 1964 [5] and in 1967 [6] and proved their importance in medicinal and technical applications. Structurally, simple 5-amino-1-tertbutylpyrazole-4-carboxamide **I** was found to inhibit p56 Lck [7]. The simple N-phenyl amide of 5-amino-1,3-dimethylpyrazole-4-carboxylic acid **II** has been shown to exhibit antifungal activity [8]. The 5-amino-1-(2,6-dichloro-4-trifluoromethyl) phenyl)-4-(3-ethoxyphenyl)-3-methyl thiopyrazole has been described as a potent GABA (γ-aminobutyric acid) inhibitor with selectivity toward insect versus mammalian receptors [9]. 5-Amino-4-benzoyl-3-methylthio-1-(2,4,6-trichlorophenyl)pyrazole **III** has been reported as a potent corti-cotrophin-releasing factor-1 (CRF-1) receptor antagonist [10]. The 5-amino-1-(4-methylphenyl) pyrazole **IV** has been tested as an NPY5 antagonist [11].

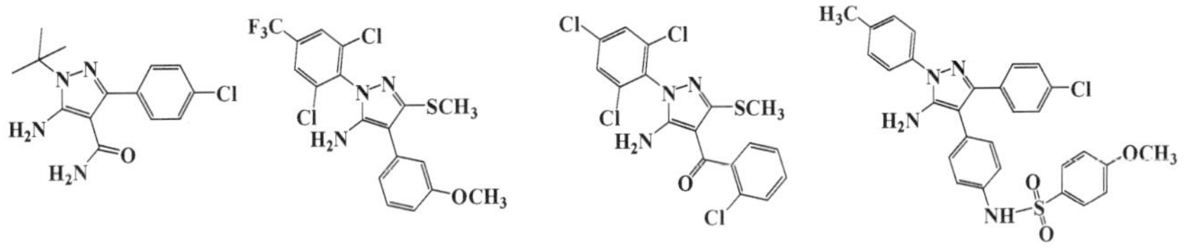

The 5-amino-1-pyrazinyl-3-carboxamidopyrazole derivatives has been recently reported as a potent antibacterial agent with a very broad spectrum [12]. Recently, the components of the mitotic machinery have been targeted in an attempt to develop novel anticancer agents. These include critical signaling kinases such as the Aurora, PLK, and the cyclin-dependent kinase (CDK). The compound (AZD1152) is the first Aurora-B selective inhibitor to enter the clinical trials [13].

2. Results and discussion

The synthesis of 3-dimethyl-2-benzoyl propenenitriles **1(a-b)** is the vital key intermediate for the synthesis of various nitrogen heterocycles, such as pyrazole and pyrimidine derivatives. The literature reports suggest that 1,3,4-trisubstituted pyrazole derivatives are important compounds in the preparation of 1,5-diphenylpyrazole nonnucleoside derivatives, which are used as HIV-1 nonnucleoside reverse transcriptose inhibitors [15]. Similarly, 4-cyano pyrazole derivatives showed significant biological activity by inhibiting alcohol dehydrogenase [16]. They also produce skeletal muscle relaxation on administration to animals [17].

In the literature, several methods have been reported for the synthesis of 5-amino pyrazole derivatives. Hasseneen and coworkers [18] have prepared pyrazole derivatives by the reaction of nitrile imine with fumaronitrile. Jachak and co-workers [19] also reported the synthesis of 4-cyano pyrazole derivatives by starting with cyanoacetaldehyde, DMF-DMA (*N,N*-Dimethylformamide dimethyl aceta) and hydrazines.

Recently, David Tupper [20] has reported the synthesis of 4-cyano pyrazole derivatives by starting with compounds similar to **1a.** These workers have prepared 4-cyano pyrazole derivatives along with 5-amino pyrazole derivative by refluxing 3-dimethylamino-2-benzoyl-propenenitrile **1a** with phenyl hydrazine or hydrazine in ethanol. However, the product was always a mixture of 4-cyano and 5-aminopyrazole derivatives. These workers have separated the mixture of pyrazoles by column chromatography and observed that the reaction of hydrazine or phenyl hydrazine took place with 3-dimethylamino-2-aroyl-propenenitrile to furnished pyrazole carbonitrile as major and aminopyrazoles as minor products.

Herein, the new route for the synthesis of 4-cyano pyrazole and 5-amino pyrazole derivatives has been described. It was demonstrated that the new procedure for the synthesis of 4-cyano and 5-aminopyrazole derivatives gave good yield. Also it was observed that treatment of **1** with hydrazine (or substituted hydrazine) in acidic medium gave 1,3-disubstituted 4-cyano-pryrazole derivatives **3**. Herein, the new route for the synthesis of 4-cyano pyrazole and 5-amino pyrazole derivatives has been described. It was demonstrated that the new procedure for the synthesis of 4-cyano and 5-aminopyrazole derivatives gave good yield. Also it was observed that treatment of **1** with hydrazine (or substituted hydrazine) in acidic medium gave 1,3-disubstituted 4-cyanopryrazole derivatives **3**. Tuper, Bray and his co-workers [20] reported that the 1,5-disubstituted–4-cyanopyrazole was obtained when compound **1** was refluxed in ethanol with hydrazine (or phenyl hydrazine).

2.1. Section I: Synthesis of 4-cyano pyrazole derivatives, 3a-i

Different methods were used for the synthesis of 4-cyano and 5-amino pyrazole derivatives. Tuper and Bray [20] performed these reactions without acid and base. Our observation was different from their studies.

The reactions of hydrazine or phenyl hydrazine with compound **2** in ethanol and catalytic amount of conc. HCl furnished 4-cyano pyrazole derivative **3(a-i)** as a single product (Experiment No. 1).

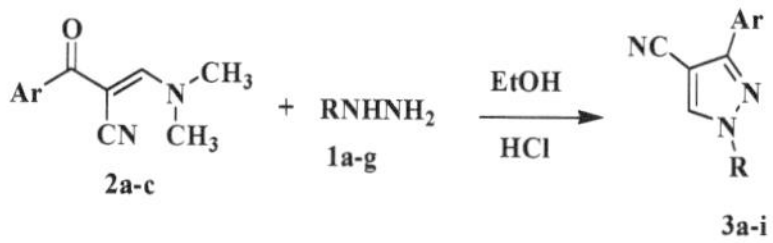

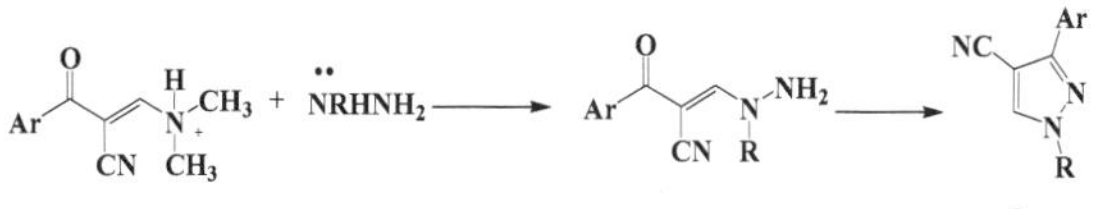

3: a, Ar= Ph, R=Ph; b, Ar=Ph, R=p-CH3Ph, c, Ar=Ph, R=p-CPh; d; Ar=Ph, R=p-NO2Ph; e, Ar=Ph, R=OCH$_3$Ph; f, Ar=Ph, R=H; g, Ar=Ph, R= CH$_2$CH$_2$OH; h, Ar=P-BrPh R= CH$_2$CH$_2$OH, I, Ar=P-BrPh R=Ph

The formation of 4-cyano pyrazole derivatives **3** can be rationalized as the acid protonated nitrogen of dimethylamino group and was replaced by hydrazine and then NH$_2$ of the hydrazine condenses with carbonyl carbon to form pyrazole ring. When the condensation of 3-dimethylamino-2-bezoyl-propenenitrile **1a** and N-methyl ester of hydrazine was carried out, the ester group has been hydrolyzed and decarboxylated to give 1H-pyrazole derivative **3f**.

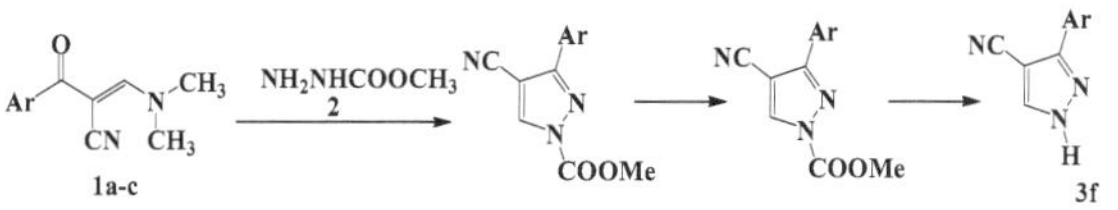

Sr. No.	Name of the compound	Solvent	N^1-R	C$_3$-Ar	C$_5$-H
1.	1,3-diphenyl-1H-pyrazole-4-carbonitrile, **3a**	Lit [22]			
2.	3-phenyl-1-p-tolyl-1H-pyrazole-4-carbonitrile, **3b**	DMSO-d$_6$	2.35, s, 3H, CH3, 7.10-7.60 m, 4H, Ar-H	7.10-7.60, m, 5H, Ar-H	8.42, s
3.	1-(4-cholorphenyl)-3-phenyl-1H-pyrazole-4-carbonitrile **3c**	DMSO-d$_6$	d ,7.60, d,4H, Ar-H	7.16-7.50, m,Ar-H	8.45, s
4.	1-(4-nitrophenyl)-3-phenyl-1H-pyrazole-4-carbonitrile, **3d**	Lit [19]			
5.	1-(4-methoxyphenyl)-3-phenyl-1H-pyrazole-4-carbonitrile, **3e**	DMSO-d$_6$	3.75 s CH3 6.90-5.77 m , 4H, Ar-H	6.90- 5.77, m,Ar-H	8.37, s
6.	3-phenyl-1H-pyrazole-4-carbonitrile, **3f**	DMSO-d$_6$	11.52, s, NH	7.40- 7.95 m 5H, Ar-H	8.00, s
7.	1-(2-hydroxyethyl)-3-phenyl-1H-pyrazole-4-carbonitrile **3g**	CDCl$_3$	3.67, t, 2H, CH$_2$,4.25 ,t 2H, CH$_2$	7.65, m, 5H, Ar-H	8.08, s
8.	3-(4-bromophenyl)-1-(2-hydroxyethyl)-1H-pyrazole-4-carbonitrile, **3h**	CDCl$_3$	3.67, t, 2H, CH$_2$ 4.25 ,t 2H, CH$_2$	7.42 & 7.65 d, 4H Ar-8H	7.95, s

Table 1. NMR of 4-cyano pyrazole 3a-j

These compounds were characterized by IR, ^{1}H NMR (Table No. 1). The IR of **3h** ($R=CH_2CH_2OH$) showed strong absorption at 2231 cm^{-1}due to CN and 3493 cm^{-1} for OH. The ^{1}H NMR in $CDCl_3$ of this compound showed clear triplet at δ 4.083 and 4.22 with J = 9.3 Hz. The aromatic protons showed para substituted pattern at δ 7.68 and 7.70 as doublet with coupling constant J=8.4 Hz. The C_5-H appeared as a sharp singlet at δ 7.89.

2.2. Section II: Synthesis of 5-Amino-4-aryl-substituted pyrazole derivatives, 4a-f

Compounds **1** and hydrazine or substituted hydrazine when refluxed in ethanol in the presence of triethylamine furnished 5-amino pyrazole derivatives **4(a-f)** in good yields (Experiment No. 2). This observation was again contradictory with Tupper's work [6]. These workers observed that when hydrazine and compound **1a** were refluxed with hydrazine or phenyl hydrazine in ethanol yielded the mixture of 4-cyano pyrazole and 5-amino pyrazole derivatives in 45–85% and 10–35% respectively. But it was observed that when base is used as a catalyst, the reaction completed within 1–2 hours, and 5-amino pyrazole derivative is only the product obtained. In this reaction, the condensation occurs by replacement of dimethylamino group and the ring closure reaction because of the attack of hydrazine moiety on nitrile function. The mechanism can be given as below.

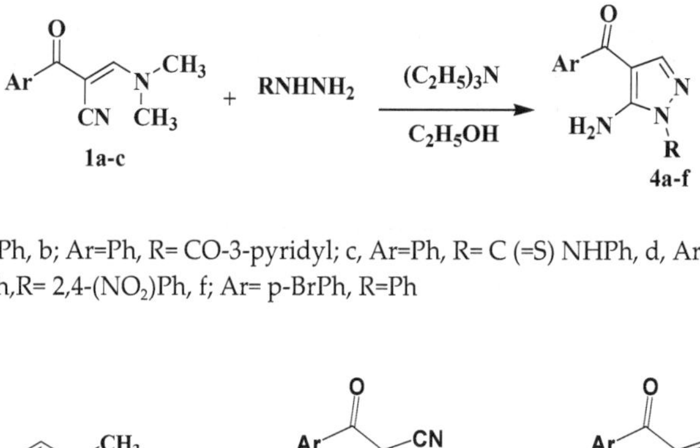

4: a, Ar=Ph, R-Ph, b; Ar=Ph, R= CO-3-pyridyl; c, Ar=Ph, R= C (=S) NHPh, d, Ar=Ph, R= CO (p-ClPh); e; Ar=Ph,R= 2,4-(NO_2)Ph, f; Ar= p-BrPh, R=Ph

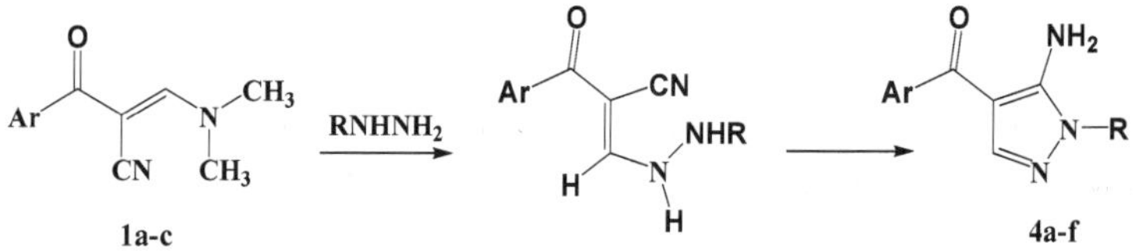

Here the other product 4-cyano pyrazole was not formed in the basic medium. The benzoyl carbonyl is less reactive, and there is no chance for the condensation of hydrazine with it. The ^{1}H NMR spectra (Table No. 2), IR of the compound **4(a-f)** characterizes all these 5-aminopyrazole derivatives. The elemental analysis was in agreement with the proposed structure. IR spectra **4f** show absorption bands at 3370 and 3320 cm^{-1} due to NH_2 group and at 1748 cm^{-1} due to the presence of carbonyl group. The ^{1}H NMR of **4f** in $CDCl_3$ showed that the NH_2 split into two singlets at δ 7.57 and 7.76 exchangeable with D_2O. The 4-aromatic p-substituted protons appears at 7.63 and 7.69 δ as doublet with J = 8 Hz. The 5-aromatic protons of the phenyl ring showed multiplet at δ 7.55–7.77, and the C_3-H appears as a singlet at δ 7.76. Thus the cyclization reaction provided synthesis for 4-cyano pyrazole and 5-amino pyrazole derivatives without a mixture of these two. The time required for the cyclization is also between 1 and 3 hours as compare to 2–18 hours as reported by Tupper and Bray [20].

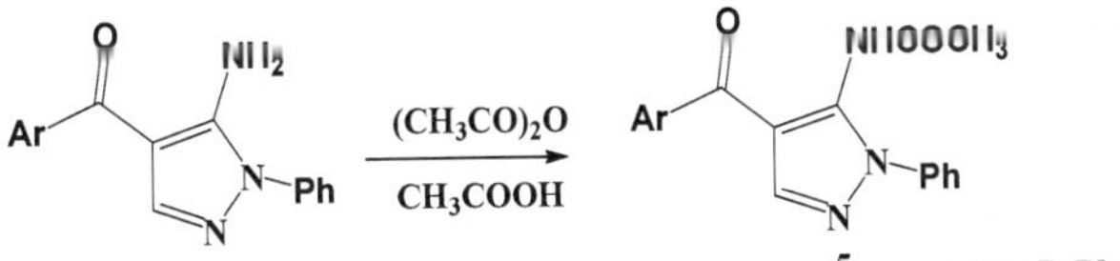

4e 5 Ar= p-BrPh

The presence of NH_2 in 5-amino pyrazole **4(a-f)** was confirmed by the formation of acetyl derivative. Thus compound **4e** on refluxing in acetic acid and acetic anhydride furnished acetyl derivative **5**.. The structure of **5** was characterized by IR and ^{1}H NMR which is given in experimental part.

2.2.1. *Deamination of 5-aminopyrazole derivatives*

In the literature, the amino group in the pyrazole system can be removed by the method explained by Nishiwaki et al [23] and Doyle et al [21]. Doyle and his coworkers have performed the reductive deamination involving arylamines. Kornblum suggested that the aromatic primary amine group was diazotized and replaced by hydrogen donor [22].

$$ArNH_2 \longrightarrow Ar\text{-}N{=}N\text{-}OR + H_2O$$

$$Ar\text{-}N{=}N\text{-}OR \longrightarrow Ar\text{-}N_2^{\cdot} + {}^{\cdot}OR$$

$$Ar\text{-}N_2^{\cdot} \longrightarrow Ar^{\cdot} + N_2$$

$$Ar^{\cdot} + Sol\text{-}H \longrightarrow Ar\text{-}H + Sol^{\cdot}$$

Thus the amino group in compounds **4a, d, e** in pyrazole on treatment with isopentylnitrile in DMF furnished deaminated pyrazole derivative **6a-c** in good yields.(ExperimentNo. 3).

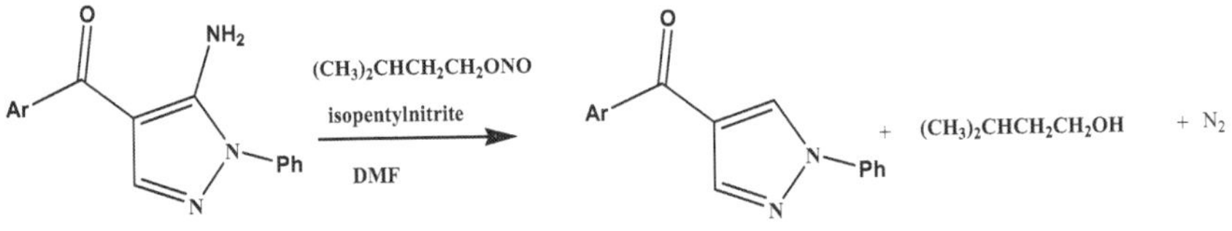

4a, d, e 6a-c

Compound 6**a-c** was characterized by ^{1}H NMR, IR, and elemental analysis. The IR spectra did not show peak at δ 3370 and 3320 cm^{-1} for NH_2 group, and the increase in carbonyl absorption from 1690 to 1720 cm–1 was due to the free carbonyl group that indicated the loss of NH_2 group. The ^{1}H NMR of **6c,** R=Ph in $CDCl_3$ showed singlet for C_3-H and C_5-H at δ 8.12 and δ 8.34 as it was expected. The four aromatic protons showed para substituted pattern at δ 7.79, 7.77 as two doublets J = 8 Hz and five protons of phenyl ring showed multiplet at δ 7.22–7.75. After deamination, the product containing carbonyl function was characterized by the formation of 2,4-DNP derivatives.

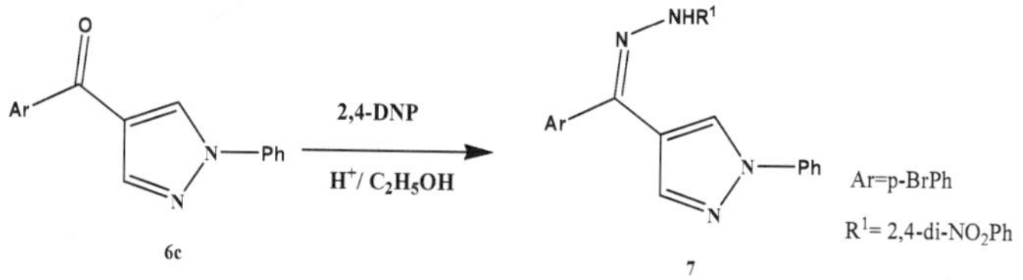

Sr. No.	Name of the compound	Solvent	N^1-R	C_4-COAr	C_3-H	C_5-NH_2
1	(5-Amino-1-phenyl-1H-pyrazole-4-yl)phenyl)methanone, **4a**	DMSO-d_6	7.40-7.75, m 5H Ar-H	7.407.75,5H, m, Ar-H	7.86, s	7.40 &7.78 s
2	(5-Amino-1-(3-pyridylcarboxy)-1H-pyrazole-4-yl)phenyl) methanone, **4b**	DMSO-d_6	8.15-8.60, m, 4H	7.40-7.75, m 5H, Ar-H	7.97, s	8.80 & 9.15, s
3	(5-Amino-1-(phenylsemicarbazide)-1H-pyrazole-4-yl) phenyl) methanone, **4c**	DMSO-d_6	7.28-7.93, m, 5H 12.05, bs, NH	7.28-7.94, m,	8.04, s	9.2 & 11.82 s
4	(5-Amino-1-(4-chlorobenzene))-1H-pyrazole-4-yl) phenyl) methanone, **4d**	DMSO-d_6	7.25-7.78 m, 4H	7.25-7.78, m, 5H	7.92, s	8.25 & 11.75 s
5	(5-Amino-1-(2,4-dinitrophenyl))-1H-pyrazole-4-yl) phenyl) methanone, **4e**	$CDCl_3$	8.23, 8.45, d & 9.23, s Ar-H	7.28 &7.65 d 4H, Ar-H	8.02, s	8.30 & 11.80, s
6	(5-Amino-1-(4-bromobenzene))-1H-pyrazole-4-yl) phenyl) methanone, **4f**	$CDCl_3$	7.02-7.56, m, 5H, Ar-H	7.26 & 7.63 d, 4H, ar-H	7.92 s	7.26 & 7.63 s, peak lost in D_2O

Table 2. NMR of 1-phenyl-4-benzoyl-5-aminopyrazole, 4a-f chemical shift in δ

Thus compound **6c** on treatment with 2,4-dinitrophenylhydrazine in acidic medium furnished the hydrazone derivative **7** and supported the presence of carbonyl group in compound **6c.** (Experiment No. 5). The ^{1}H NMR clearly indicated the singlet at δ 11.15. for NH protons and 3 hydrogen of phenyl group containing nitro group clearly observed at δ 8.09, 8.22, and 9.05.

3. Conclusion

The reaction of aroylpropenenitrile and substituted hyrazines in the presence of acid catalyst yielded 4-cyano pyrazoles and the same reaction in basic medium yielded 5-amino pyrazole derivatives as signal product in good yields.

To an equimolar solution (0.01 mol) of **1(a-b)** and substituted hydrazine **2(a-i)**, in ethanol (30 ml), concentrated hydrochloric acid (0.2 ml) was added, and the reaction mixture was refluxed for the time shown below. The solvent was removed in vacuo to get the residue of **3(a-j)**, which was recrystallized from the proper solvent.

1,3-Diphenyl-4cyanopyrazoles, 3a

Heating under refluxed for 3.5 hours, yield 65%, recrystallized from ethanol, m.p. 134°C (lit. [21] m.p. 135°C).

1-p-Toloyl-3-phenyl-4-cyanopyrazole, 3b

Heating under refluxed for 3 hours, yield 68%, recrystallized from ethanol, m.p. 123°C, IR(KBr):2230 and 1520 cm^{-1}.

1-(p-Chlorophenyl)-3-phenyl-4-cyanopyrazole, 3c

Heating under refluxed for 3 hours, yield 75%, recrystallized from ethanol, m.p. 141°C. IR (KBr): 2240, 1505 cm^{-1}.

1-(4-nitrophenyl)-3-phenyl-1H-pyrazole-4-carbonitrile,3d

Heating under refluxed for 3.5 hours, yield 70%, recrystallized from ethanol, m.p. 223°C, (lit. [19] m.p. 225°C).

1-(4-methoxyphenyl)-3-phenyl-1H-pyrazole-4-carbonitrile,3e

Heating under refluxed for 1 hour, yield 75%, recrystallized from ethanol, m.p. 125°C. IR (KBr): 2228 and 1510 cm^{-1}.

3-phenyl-1H-pyrazole-4-carbonitrile, 3f

Heating at 60°C for 6 hours, yield 60%, recrystallized from ethanol, m.p. 131°C (lit. [19] m.p. 134°C. IR (KBr): 3150, 2960, 2240, and 1510 cm^{-1}.

1-(2-hydroxyethyl)-3-phenyl-1H-pyrazole-4-carbonitrile, 3g

Heating under refluxed for 2.5 hours, yield 65%, recrystallized from ethanol, m.p. 106°C. IR (KBr): 2228, 1510 cm^{-1}.

3-(4-bromophenyl)-1-(2-hydroxyethyl)-1H-pyrazole-4-carbonitrile, 3h

Heating under refluxed for 2.5 hours, yield 63%, recrystallized from methanol, m.p. 135°C. IR (KBr): 2231, 1563, and 1533 cm^{-1}.

3-(4-Bromophenyl)-1-phenyl-1H-pyrazole-4-carbonitrile, 3i

Heating under refluxed for 2 hours, yield 68%, recrystallized from methanol, m.p. 210°C. IR (KBr): 2210, 1600, 1580, and 1533 cm^{-1}.

3.1. Experiment 2

Synthesis of 1-substituted-4-benzoyl-5-aminopyrazoles, 4(a-h)

To an equimolar solution of **1a** or **1b** (0.01 mol), substituted hydrazines, **2a, j-n** in ethanol (30 ml) was taken in a reaction flask. Triethlyamine (0.2 ml) was added, and the reaction mixture was heated under reflux for the time shown below. The solvent removed in vacuo and the product obtained was filtered, recrystallized from the solvent shown for individual compound.

(5-Amino-1-phenyl-1H-pyrazole-4-yl)phenyl)methanone, 4a

Heating at 65°C for 1 hour, yield 65%, m.p. 178°C. IR (KBr): 3380, 3275, 1620, and 1540 cm^{-1}.

(5-Amino-1-(3-pyridylcarboxy)-1H-pyrazole-4-yl)phenyl) methanone, 4b

Heating under reflux for 1.5 hours, yield 45%, m.p. 149°C. IR (KBr): 3460, 3320, 3050, 1705, 1695, and 1630 cm^{-1}.

5-Amino-1-(phenylsemicarbazide)-1H-pyrazole-4-yl) phenyl) methanone, 4c

Heating under reflux for 1.5 hours, yield 450%, m.p. 127°C. IR (KBr): 3380, 3300, 3140, 1640, 1600, and 1550 cm^{-1}.

(5-Amino-1-(4-chlorobenzene))-1H-pyrazole-4-yl) phenyl) methanone, 4d

Heating under reflux for 1.5 hours, yield 45%, m.p. 199°C, recrystallized from ethanol. IR (KBr): 3370, 3320, 3040, 1690, 1630, 1590, and 1550 cm^{-1}.

(5-Amino-1-(2,4-dinitrophenyl))-1H-pyrazole-4-yl) phenyl) methanone, 4e

Heating under reflux for 2 hours, yield 60%, m.p. 217°C, recrystallized from ethanol. IR (KBr): 3443, 3221, 3050, 2922, 1741, 1631, 1605, and 1550 cm^{-1}.

(5-Amino-1-(4-bromobenzene))-1H-pyrazole-4-yl) phenyl) methanone, 4f

Heating under reflux for 1 hour, yield 50%, m.p. 186°C, recrystallized from ethanol. IR (KBr): 3370, 3320, 3040, 1690, 1630, 1590, and 1550 cm^{-1}.

3.2. Experiment 3

Deamination of 5-amonopyrazole derivatives: Preparation of 1-Substituted-4-benzoyl pyrazole, 6(a-d)

To a solution of **3a, d, or f** (0.01 mol) in anhydrous dimethylformamide (5 ml) maintained at 60–65°C, isopentylnitrite (0.01 5 mol) in anhydrous DMF (3 ml) was added over 10 minute. The mixture was stirred for 30 minutes. The solvent was removed under reduced pressure to get solid. The solid obtained was filtered washed with petroleum ether and recrystallized from proper solvent.

1-Phenyl-4-benzoylpyrazole, 6a

M.p. 124°C (lit [22]. m.p. 123–124°C); recrystallized from ethanol **1-(p-Chlorophenylcomicar-bazole)-4-benzoylpyrazole, 6b** m.p. 235°C, recrystallized from ethanol, yield 65%. IR(KBr): 3320, 1660, 1591, 1573, and 1490 cm^{-1}. ^1H NMR (DMSO-d_6) δ: 7.25–7.78 (m, 10H, Ar-H); 7.92 (s, 1H, C_3-H); 12.12 (bs 1H, NH). **1-Phenyl-4-benzoylpyrazole, 6c** yield 76%, and m.p. 198–199°C. Recrystallized from methyl alcohol. IR (KBr): 1720, 1626, 1582, and 1562 cm^{-1}. ^1H NMR ($CDCl_3$) δ: 7.22–7.75 (m, 5H, Ar-H); 7.63 and 7.77 (d 4H, Ar-H); 8.13(s, 1H, C3-H); 8.43(s 1H C-5-H).

3.3. Experiment 4

Synthesis of 2,4-dinitrophenylhydrazone derivative of 1-phenyl-4-benzoylpyrazole, 7. In the mixture 1-phenyl-4-benzoylpyrazole (0.002 mol, 0.642 gm), 2,4-dinitro phenyl hydrazine (0.002 mol, 0.396 gm) in ethyl alcohol (20 ml), concentrated sulfuric acid (0.2 ml) was added. The reaction mixture was refluxed for 3 hours. The solvent was removed and solid obtained was filtered, washed with ethanol and recrystallized from ethanol: DMF (2:8), yield 400 mg, 76%, m.p. 220°C. IR (KBr): 3340, 1626, 1590, 15550, and 1480 cm^{-1}. ^1H NMR ($CDCl_3$) δ: 7.25-7.51(m, 5H, Ar-H); 7.70 and 7.85 (d 4H, J=8.4 Hz, Ar-H); 7.95(s, 1H, C_3-H); 8.04(s 1H C_5-H); 8.25 & 8.40 (d 2H J=8.4 HzAr-H); 9.15(s, 1H Ar-H).

Author details

Raghunath Toche[1,2*]

Address all correspondence to: raghunath_toche@rediffmail.com

1 Department of Chemistry, KRT Arts, BH Com and AM Science College, Nashik, Pune, India

2 University of Pune, Pune, India

References

[1] Elguero, J. In *Comprehensive Heterocyclic Chemistry*; Katritzky, A. R.; Rees, C. W., eds.; Pergamon Press: Oxford, 1984; Vol. 5, pp. 167–303. doi:10.1016/B978-008096519-2.00072-2.

[2] Elguero, J. In *Comprehensive Heterocyclic Chemistry II;* Katritzky, A. R., Rees, C. W., and Scriven, E. F. V., eds.; Pergamon Press: Oxford, 1996; Vol. 3, pp. 1–75. doi: 10.1016/ B978-008096518-5.00059-9.

[3] Kost, A. N. and Grandberg, I. I. In *Advances in Heterocyclic Chemistry;* Katritzky, A. R. and Boulton, A. J., eds.; Academic Press: New York, 1966; Vol. 6, pp. 347 ff.

[4] Lee, K. Y., Kim, J. M., and Kim, J. N. *Tetrahedron Lett.* 2003, 44, 6737–6740. doi:10.1016/ S0040-4039(03)01648-4.

[5] Wiley, R. H. and Wiley, P. *Pyrazolones, Pyrazolidones and Derivatives;* John Wiley and Sons: New York, 1964.

[6] Behr, L. C., Fusco, R., and Jarboe, C. H. In *The Chemistry of Heterocyclic Compounds, Pyrazoles, Pyrazolines, Pyrazolidines, Indazoles and Condensed Rings;* Weissberger, A., ed.; Interscience Publishers: New York, 1967.

[7] David, D. P., Martin, D. J., and Charles, M. D. F. 5-Aminopyrazoles useful as selective inhibitors of the protein tyrosine kinase P56ick. WO 9740019 (A1), Nov. 30, 1997.

[8] Kordik, C. P., Luo, C., Zanoni, B. C., Lovenberg, T. W., Wilson, S. J., Vaidya, A. H., Crooke, J. J., Rosenthal, D. I., and Reitz, A. B. *Bioorg. Med. Chem. Lett.* 2001, 11, 2287–2290. doi:10.1016/S0960-894X(01)00449-8

[9] Nakazato, A. and Okuyama, S. *Drugs Future* 1999, 24, 1089–1098. doi:10.1358/dof. 1999.024.10.665576.

[10] Meegalla, S. K., Doller, D., Sha, D., Soll, R., Wisnewski, N., Silver, G. M., and Dhanoa, D. *Bioorg. Med. Chem. Lett.* 2004, 14, 4949–4953. doi:10.1016/j.bmcl.2004.07.033.

[11] Huppatz, J. L. *Aust. J. Chem.* 1985, 38, 221–230. doi:10.1071/ CH9850221.

[12] Shamroukh, A. H., Rashad, A. E., and Sayed, H. H. *Phosphorus, Sulfur Silicon Relat. Elem.* 2005, 180, 2347–2360. doi:10.1080/ 104265090921074.

[13] Carter, T. A., Wodicka, L. M., Shah, N. P., Velasco, A. M., Fabian, M. A., Treiber, D. K., Milanov, Z. V., Atteridge, C. E., Biggs, W. H., Edeen, P. T., Floyd, M., Ford, J. M., Grotzfeld, R. M., Herrgard, S., Insko, D. E., Mehta, S. A., Patel, H. K., Pao, W., Sawyers, C. L., Varmus, H., Zarrinkar, P. P., and Lockhart, D. J. *Proc. Natl. Acad. Sci. U. S. A.* 2005, 102, 11011–11016. doi:10.1073/pnas.0504952102.

[14] Ranjana Aggarwal, Vinod Kumar, Rajiv Kumar, and Shiv P. Singh *Beilstein J. Org. Chem.* 2011, 7, 179–197. doi:10.3762/bjoc.7.25.

[15] M. J. Genin, C. Biles, B. J. Keiser, S.M. Poppe, S.M. Swaney, W.G. Tarplay, V. Yagi, and D. L. Ramero; J. Med. Chem. 2000, 43, 1034–1040.

[16] Coenell N.W., Honsch C., Kim K. H., and Heneger K., Arch. Biochem. Biophys. 1983, 227, 81–90.

[17] Harry R. Snyder, Chem Abst. 66, 46424 (1967). U.S. Pat. 3,293,261 (1966).

http://dx.doi.org/10.5772/64050

[18] H. M. Hasseneen, H. A. Edd, N m. Elwan, and A. S. Shawali, Heterocycles, 27, 2857 (1988).

[19] M. Jachak, U. Kriessman, M. Mittlebach, and H. Junek, Monatshefte Chemie, 124, 199–207 (1999).

[20] David E. Tupper and Mark R. Bray, Synthesis 337, 1996.

[21] M. P. Doyle, J. F. Dellaria, Jr. B. Siegfried, and S.W. Bishop, J. Org. Chem. 42(22), 3494 (1977).

[22] N. Kornblum, Org. React. 2, 262 (1944).

[23] T. Nishiwaki, F. Fusko, and Emil Minamisono, J.C.S. Perkin I, 1871 (1974).